ENGINEERING TRADE SPECIALIZATION OF SWEDEN AND OTHER INDUSTRIAL COUNTRIES

STUDIES IN INTERNATIONAL ECONOMICS

Volume 6

NORTH-HOLLAND PUBLISHING COMPANY
AMSTERDAM · NEW YORK · OXFORD

ENGINEERING TRADE SPECIALIZATION OF SWEDEN AND OTHER INDUSTRIAL COUNTRIES

A Study of Trade Adjustment Mechanisms of Factor Proportions Theory

LENNART OHLSSON
Economic Research Institute
Stockholm School of Economics
Stockholm, Sweden

Industriens Utredningsinstitut

1980

NORTH-HOLLAND PUBLISHING COMPANY
AMSTERDAM • NEW YORK • OXFORD

ISBN: 0 444 86114 9

Publishers:
NORTH-HOLLAND PUBLISHING COMPANY
AMSTERDAM • NEW YORK • OXFORD

Sole distributors for the U.S.A. and Canada:
ELSEVIER NORTH-HOLLAND, INC.
52 VANDERBILT AVENUE
NEW YORK, N.Y. 10017

Library of Congress Cataloging in Publication Data

Ohlsson, Lennart, 1942-
Engineering trade specialization of Sweden and other industrial countries.

(Studies in international economics ; v. 6)
Bibliography: p.
Includes index.
1. Engineering--Sweden. 2. Engineering.
3. Interindustry economics. I. Title.
HD9680.S82034 339.2'3 80-21628
ISBN 0-444-86114-9

PRINTED IN THE NETHERLANDS

INTRODUCTION TO THE SERIES

This series is intended to embrace all aspects of international economic analysis: theoretical, empirical and econometric. It will equally encompass contributions to pure and monetary theory.

The series will include the publication of collected essays, conference proceedings of exceptional professional quality and interest, and edited volumes addressed to specific phenomena of importance in the analysis of international economics. It will particularly encourage, however, the publication of hitherto unpublished studies which represent fresh and valuable contributions to the analysis of international economic problems.

The Journal of International Economics now provides an excellent outlet for original scientific work in international economics in article length form. For pamphlet length contributions, the different series published by the *International Finance Section* at Princeton have long been of exceptional value, even though they have not generally extended beyond international monetary issues. It is the hope of the editors that the launching of this series, *Studies in International Economics*, will now fill the lacuna in the systematic publication of monograph and volume length research output of international economists, and succeed in doing so consistently with the high quality that the promotion of scientific study of international economic issues requires.

THE EDITORS

Bo Carlsson and Lars Lundberg for their expert advice and Olle Renck for his careful examination and criticism of the manuscripts. Indirectly, via early comments on journal papers, my work was both improved and encouraged by helpful suggestions from especially Gary C. Hufbauer, Harry G. Johnson and Donald B. Keesing. Also Herbert G. Grubel, John C. Hause, Robert E. Lipsey, Peter J. Lloyd, Harry Lütjohann and anonymous referees of several journals have been very helpful.

Throughout the work with the English book John S. Chipman provided encouragement. His careful editorial advice certainly meant substantial improvements of many weak points in the English manuscript. William H. Branson's helpful comments on material partly available only in Swedish is gratefully acknowledged. Gunnar Du Rietz, my co-author of chapter 7, was kind enough to allow a publication of our joint work in this book. We both have reasons to be grateful to John C. Hause and to Richard E. Caves for helpful suggestions on earlier versions of that chapter.

Stanley Noval examined my English and found many errors. I do hope that possible remaining ones are few and minor.

Finally, some advice to the potential reader. From a policy point of view, the most interesting and powerful results are presented in chapters 6, 7, 9 and 10.

Lennart Ohlsson

TABLE OF CONTENTS

LIST OF TABLES

LIST OF FIGURES

Chapter 1

INTRODUCTION

1.1 Background and aim of study

The liberalization of trade since the 1940's involved the breakdown of major non-tariff barriers in the late 1940's and 1950's and of tariff barriers in the 1960's and early 1970's. During this time natural trade impediments such as transport and information costs appear to have fallen relative to other costs of production. For Sweden as well as other industrial countries the international dependence increased in many ways as a consequence.

One way in which international dependence increased was that the structure of industrial production grew less associated with the composition of industrial goods in domestic consumption. This increase in the international division of labour meant that Swedish production costs became gradually less important as a determinant of the internal commodity price structure. Swedish producers had to face more exogenously given domestic and world market prices. Being a small and increasingly open economy, the allocation of resources within the tradable goods sector became more and more determined by the international competitiveness of domestic producers.

During this period of trade liberalization Sweden's rôle as a supplier of raw materials and raw material based goods diminished. Instead it became a large producer and exporter of engineering products (cf Ohlsson [1969] chapter 1). By the end of the 1960's Sweden's share of engineering exports in total exports was among the largest of all OECD countries. This development occurred during a period in which engineering goods grew rapidly in proportion to the manufacturing production of industrial countries. Their rôle in international trade became even more pronounced since many engineering products are highly tradable.

With this background the present study was first set out to investigate the international competitiveness of various *types* of Swedish engineering products and the structural development of foreign trade in those products. Due to the lack of explanatory power of preliminary analytical approaches the task changed to explaining the patterns of Swedish engineering trade with the neo-Heckscher-Ohlin or modern factor proportions theory. In the English edition of the study the methodological aspects have been emphasized even more.

The purpose of the present study is three-fold. First, it presents the results of an analysis of the patterns of engineering specialization in Sweden and other industrial countries. These results should clarify the nature of some adjustment mechanisms associated with earlier changes in trade policy.

The second aim of the English edition is to make available results which might stimulate further theoretical development in areas of special interest to future empirical applications. The third aim is to present new methodologies which facilitate the empirical study of trade specialization patterns in technologically heterogeneous industries.

1.2 Theory and methodology

Among the many possible determinants of the trade pattern of a country only certain product- or process-specific ones were selected for the analysis. The study concentrates more on possible macroeconomic determinants behind the product composition of engineering trade than on microeconomic ones, such as the characteristics of individual firms, their technological change, location, etc.

Given this focus of interest the theoretical framework was built upon trade theories which might be classified as comparative advantage theories. A comparative advantage theory is narrowly defined as a theory which derives the competitiveness of each country in free trade from a comparison of the commodity price structures of the countries in assumed states of autarky.

According to this narrow and abstract definition two sets of theories qualify as comparative advantage theories: 1) theories based on the classical Ricardo-Torrens theory of comparative costs and 2) certain factor proportions theories based upon the original works of the Swedish economists Eli Heckscher and Bertil Ohlin. Now, if instead a broader and necessarily looser definition is chosen in which the autarky situation is abandoned classes of trade theories other than the two mentioned might be labelled comparative advantage theories.

The present study investigates the rôle of macroeconomic determinants of the commodity trade pattern. Hence the so-called product-cycle theory can be regarded as a possible theoretical framework along with the abovementioned theories. The product cycle theory was developed by Raymond Vernon and Seev Hirsch under the influence of the work of another Swedish economist, Staffan Burenstam Linder. Both factor proportions and product cycle theories assign national characteristics a decisive influence on the commodity trade pattern. As for the classical comparative cost theory it is less clear what grounds there are for the assumption that a given country is more efficient in the production of certain commodities. Hence, the choice of a theoretical framework was between the factor proportions and product cycle theories.

Being better developed formally, more thoroughly analysed and easier to apply empirically, the factor proportions approach was finally selected as our frame of reference. However, this choice did not preclude using some elements of the product cycle theory to generate certain testable hypotheses.

Among theories being labelled factor proportions theories there are some which deviate substantially from the assumptions of the original Heckscher-Ohlin-Samuelson (H-O-S) model. A fundamental common characteristic is that all investigate how differences in the factor endowments (or prices) of countries and factor requirements of commodities determine *commodity* trade patterns and factor prices. Thus there are factor proportions theories underlining the importance of for instance technical change or demand and there are static, comparative static and dynamic factor proportions theories. The use of such a rich body of theories has two advantages. One is that one does not *a priori* exclude as many possible determinants from consideration. A second advantage is that it may be possible to combine a factor proportions approach, here based on some multicommodity-multifactor-two country static models with elements from other trade theories. A disadvantage is that the design of the empirical analysis may become rather arbitrary with respect to why some tests have been preferred to others.

The main body of economic theory rests on assumptions which abstract from aggregation problems connected with all empirical applications. The factor proportions theories are no exception in this respect. The fact that the engineering industry is very heterogeneous deviating substantially from the traditional assumption that each commodity is homogeneous and produced in separate production processes *at aggregate industry levels* emphasized the need for explicit recognition of certain aggregation problems. However, while highly disaggregated data better fulfill the requirement of homogeneous products, it introduces other problems connected with for instance joint production and consumption, multiproduct plants, multiplant firms and multinational firms etc. Analytically, an engineering product is often difficult to define from the point of view that each producer has not the same unique combination of activities. Especially in the engineering industry there are many alternative ways to use excess capacity of a production factor and to avoid problems of too little capacity of other factors.

These characteristics of engineering production may, however, have several advantages from the point of view of the working of the adjustment mechanisms of the factor proportions theory. First of all, economies of scale that exist for a given manufacturer of, say, cars may not preclude efficient smallscale production of other cars if organized with a different set of activities. Secondly, there is a stronger potential competition due to relatively low barriers of entry associated with the production technology.

Thirdly, the tradability of engineering products, the flexibility of the organization of production and the fact that the laws of mechanics are less restrictive and require less physical capital than the laws of chemistry suggest that the adjustment to changing trade incentives may be more rapid than in other industries. On the other hand, there are other, sometimes temporary ways to adjust than those envisioned in pure trade theories.

Fourthly, the existence of multiplant producers and multinational firms does not mean *per se* that the adjustment mechanisms of factor proportions theories are less realistic. The stronger the competitive pressure in international commodity markets and in domestic factor markets, the more the within-firm-adjustment ought to follow the patterns suggested by pure trade theory.

The present study emphasizes certain aggregation problems of factor proportions theories by analysing interindustry trade at the lowest industry level as well as intra-industry and even intra-commodity group trade. In addition, the choice of specialization measures is made with due respect to the aggregation issue. Several definitions of products and processes are used in order to diminish the risk of falsely rejecting a factor proportions account because of unfortunate empirical definitions of theoretical concepts. Similar precautions were taken in the definitions of independent variables measuring the factor requirements. However, the present study reports only those approaches which were judged to be the most interesting from a methodological point of view or the most favourable for factor proportions accounts.

1.3 Definitions of the engineering industry

The present study defines the Swedish engineering industry as that part of the manufacturing industry which produces engineering products in plants located within Sweden. Hence, a definition in terms of *commodities* is chosen rather than in terms of plants or firms. In consequence, no distinction is made according to whether the production plant is owned by Swedes or not. Likewise it does not have to be an engineering firm which produces the engineering product.

This definition was chosen before the new Swedish industrial classification system (SNI = Svensk Näringsgrensindelning) based on ISIC of 1968 was available. It followed the definition of the Swedish Institute of Economic Research in terms of SITC-Revised which meant the inclusion of the following product categories: fabricated metal products (SITC 69), non-electrical machinery and appliances (71), electrical machinery and appliances (72), transport equipment (73), and certain miscellaneous fabricated metal products, namely sanitary and other metal products for construction purposes (812), scientific and technical instruments (861) and watches and clocks (864). Those product categories compare approximately to SNI division 38.

The publication of the new SNI-system in 1970 for the year 1968 allowed for extensive cross-industry analyses since the number of industries increased from about five to about forty. As a supplement to the commodity definition it was, in consequence, necessary to define the engineering industry according to industry groups. In this study the following industries were excluded from SNI division 38, which division is usually associated with the concept engineering industry (verkstadsindustri):

1. Manufacture of furniture and fixtures primarily of metal (SNI 3812)

2. Light bulb and fluorescent lamp manufacturing (SNI 38393)
3. Ship building and repairing (SNI 38411)
4. Boat building and repairing (SNI 38412)
5. Aircraft and aircraft engine manufacturing (SNI 38451)
6. Manufacture of photographic and optical goods (SNI 3852)
7. Manufacture of watches and clocks (SNI 3853)
8. All repair industries (SNI 38292, 38394, 38414, 38422 and 38452).

Some of these industries were excluded because the raw material base and thus the technology differed substantially from that of other industries. Trade with second-hand ships and difficulties with the periodicity of the output figures of the shipyards were the main reasons for excluding that industry. The aircraft industry of Sweden is a pure defense industry; both imports and exports of military aircraft are decided politically rather than commercially. There are no figures reported on the watches and clock industry. Lastly, the five repair industries are all nontradable goods industries.

After exclusion of the above industries the engineering industry contains 31 industries at the 5-digit level and 34 at the 6-digit level. Those industries constitute our cross-section at the industry level of analysis.

Table 1:1 presents figures for the engineering industries which were classified as tradable goods industries. In 1970 these 40 industries had almost 4 000 plants employing 380 000 workers or 42% of the employment in the manufacturing industry as a whole. Their share of total value added of the manufacturing industry was slightly smaller. According to the more narrow definition of the engineering industry chosen in the present study this sector employed 36% of the employees of the manufacturing industry.

According to table 1:1 no industry had a value added much larger than 5% of that of the whole engineering industry. Several shares were smaller than 1%.

Table 1:2 presents figures on the absolute and relative importance of Swedish trade in engineering goods. As defined here shares of total and manufacturing exports, respectively, were 40 and 50% and the corresponding import shares about 7 percentage points smaller.

Table 1:1

Some measures on the size of the Swedish engineering industry and its subindustries in 1970

SNI no.	Industry	Value added		No. of employees		No. of plants		Share of manual workers employed in plants with less than 200 employees
		Mill. Skr.	per cent of total engineering industry		per cent of total engineering industry		per cent of total engineering industry	
		(1)	(2)	(3)	(4)	(5)	(6)	(7)
3811	Manufacture of cutlery, hand tools and general hardware	489	2.7	7 732	2.0	147	3.8	44
3812	Manufacture of furniture and fixtures primarily of metal	212	1.2	4.666	1.2	82	2.1	46
3813	Manufacture of structural metal products	1 054	5.7	21 822	5.7	645	16.7	88
38191	Manufacture of metal containers	157	0.9	2 690	0.7	32	0.8	53
38192	Wire cloth, wire and cable manufacturing	228	1.2	4 594	1.2	51	1.3	57
38193	Nail, bolt and nut manufacturing	235	1.3	4 915	1.3	60	1.6	43
38194	Manufacture of other metal products for construction purposes	505	2.7	11 676	3.1	139	3.6	50

Cont.

SNI no.	Industry	(1)	(2)	(3)	(4)	(5)	(6)	(7)
38195	Household metal ware manufacturing	168	0.9	3 784	1.0	72	1.9	58
38199	Manufacture of other metal products	975	5.3	21 218	5.6	645	16.7	80
3821	Manufacture of engines and turbines	131	0.7	2 133	0.6	-	-	0
3822	Manufacture of agricultural machinery and equipment	427	2.3	8 511	2.2	71	1.8	29
38231	Metal-working machinery manufacturing	336	1.8	8 041	2.1	103	2.7	61
38232	Wood-working machinery manufacturing	121	0.7	2 270	0.6	33	0.9	68
38241	Pulp and paper mill machinery manufacturing	171	0.9	3 315	0.9	30	0.7	39
38242	Construction and mining machinery manufacturing	284	1.5	5 541	1.5	60	1.6	40
38249	Manufacture of industrial machinery not elsewhere classified	676	3.7	12 259	3.2	265	6.9	77
38251	Manufacture of computing equipment	112	0.6	1 242	0.3	5	0.1	2
38259	Manufacture of other office and accounting machinery	475	2.6	11 826	3.1	51	1.3	19

Cont.

SNI no.	Industry	(1)	(2)	(3)	(4)	(5)	(6)	(7)
38291	Manufacture of household appliances, except purely electrical	400	2.2	9 532	2.5	24	0.6	18
382991	Lifting and hoisting machinery manufacturing	633	3.4	12 044	3.2	143	3.7	53
382992	Liquid pump manufacturing	199	1.1	4 204	1.1	30	0.7	40
382993	Manufacture of general purpose parts of machinery	456	2.5	10 836	2.9	32	0.8	10
382999	Manufacture of other machinery and equipment	1 557	8.5	32 446	8.5	211	5.5	29
3831	Manufacture of electrical industrial machinery and apparatus	630	3.4	15 562	4.1	61	1.6	15
3832	Manufacture of radio, television and communication equipment and apparatus	1 350	7.3	33 561	8.8	98	2.5	11
3833	Manufacture of electrical appliances and housewares	232	1.3	3 996	1.1	37	1.0	33
38391	Manufacture of insulated wires and cables	382	2.1	4 301	1.1	12	0.3	20
38392	Storage battery and accumulator manufacturing	133	0.7	2 415	0.6	13	0.3	25
38393	Light bulb and fluorescent lamp manufacturing	473	2.6	1 645	0.4	22	0.6	85

Cont.

SNI no.	Industry	(1)	(2)	(3)	(4)	(5)	(6)	(7)
38399	Manufacture of other electrical equipment	275	1.5	6 806	1.8	79	2.1	52
38411	Ship building and repairing	964	5.2	25 233	6.6	48	1.2	9
38412	Boat building and repairing	85	0.5	2 150	0.6	91	2.4	95
38413	Marine engine manufacturing	101	0.5	1 442	0.4	4	0.1	45
38421	Railroad equipment manufacturing	108	0.6	3 041	0.8	5	0.1	2
38431	Motor vehicle and chassi manufacturing	1 288	7.0	21 700	5.7	14	0.4	3
38432	Manufacture of motor vehicle engines, parts and trailers	986	5.4	24 434	6.4	242	6.3	38
3844	Manufacture of motorcycles and bicycles	54	0.3	1 391	0.4	15	0.4	33
3845	Manufacture and repair of aircraft	936	5.1	16 584	4.4	34	0.9	22
3849	Manufacture of transport equipment not elsewhere classified	48	0.3	1 137	0.3	50	1.3	94
3851	Manufacture of professional and scientific, and measuring and controlling equipment, not elsewhere classified	336	1.8	7 021	1.8	104	2.7	65
Engineering industry: total		18 382	100.0	379 616	100.0	3 860	100.0	-
as per cent of manufacturing industry		38.7	-	41.8	-	28.9	-	-

Cont.

	(1)	(2)	(3)	(4)	(5)	(6)	(7)
Engineering industry							
(excl. 3812, 38393, 38411, 38412, 3845, 3849):							
total	15 664	-	328 201	-	3 533	-	-
as per cent of manufacturing industry	33.0	-	36.2	-	26.5	-	-

Source: SOS, Industri 1970, del 1.

Table 1:2

Engineering products in Swedish trade 1970

	Exports	Imports	Net exports
Total trade, in billions of Skr	35.2	36.3	-1.1
of which			
manufacturing products (SITC 5-8), in billions of Skr	27.0	26.5	0.5
of which			
engineering products (SITC 69, 7, 812, 861, 864), in billions of Skr	15.7	12.6	3.1
of which			
engineering products in the present study, in billions of Skr	13.7	11.6	2.1
All engineering products as per cent of total trade	44.8	34.7	-
All engineering products as per cent of traded manufacturing products	58.2	47.4	-
Engineering products of this study as per cent of total trade	39.1	32.1	-
Engineering products of this study as per cent of manufacturing products	50.8	43.8	-

Source: SOS, Utrikeshandel, del 2 1970.

It is important to emphasize that the use of Swedish industrial and trade statistics means that the analysis covers the activities located at the manufacturing plant while as activities located elsewhere and employing only whitecollar personnel are by definition generally excluded from the industrial sector.

1.4 Outline of the study

The present study contains ten chapters, two appendices and an additional appendix containing basic Swedish data tables. Chapters 2 and 3 are introductory in character. *Chapter 2* presents the theoretical and methodological framework of the study, while *chapter 3* discusses Sweden's factor abundance and comparative advantage[1].

Chapters 4-7 analyze the Swedish inter- and intra-industry pattern of specialization. *Chapter 4* focuses on the inter-industry pattern in 1970 while *chapter 5*[2] tests whether the results of chapter 4 can be attributed to heterogeneous industries and intra-industry trade.

Chapters 6-7 are designed to study changes in inter-industry specialization during the 1960's. *Chapter 6* covers the influence of changes in tariffs, technology, factor abundance and demand though still preserving many of the traditional assumptions of the simple Heckscher-Ohlin-Samuelson model. Some of those assumptions are abandoned in *chapter 7*, which analyses the comparative development, or trends in specialization, of permanent domestic firms, new entries, exiting firms and foreign firms. That chapter emphasizes the rôle of factor and commodity market differentials for trade and production adjustment in the medium run. Chapter 7 is jointly written with Gunnar Du Rietz[3].

Some of the results of the inter-industry analysis were found so important that an attempt to generalize them to a) lower levels of aggregation and b) other industrial countries seemed called for. That analysis is contained in chapters 8 and 9[4], which cover inter- and intra-commodity group patterns of specialization for 14 industrial countries. *Chapter 8* investigates the 1970 patterns while *chapter 9* studies the 1964-1970 specialization trend as well as the 1964 pattern.

Chapter 10 summarizes the results of the monograph. It also presents a broader interpretation of the main findings as regards both methodological and economic policy implications.

Footnotes

1. Compared to the corresponding sections in the Swedish edition the first three chapters have been substantially reduced. The reason is, of course, that the English edition is intended for a more professionally trained readership.

2. The Swedish edition contained only a summary of the findings in chapter 5 in appendix (E).

3. It was not included in the Swedish monograph.
4. Which compares to chapters 5 and 8 respectively 7 and 9 of the Swedish edition.

Chapter 2

AN INTRODUCTION INTO THE THEORETICAL FRAMEWORK

2.1 Introduction

The present study rests methodologically upon certain properties of a set of factor proportions theories, which can be called multicommodity-multifactor-two country theories for small, open economies. The nature of the input-output relationships is, because of the application to the engineering industry, assumed to approximate the conditions of so called inter-industry-flows-models, i.e. the output of one part of this industry tends to be used in many (all) other parts.

The assumption that Sweden is a small, open economy is taken to mean, first, that the country is a price taker in all its trade exposed sectors. Second, the intermediate goods used in its engineering industry are exposed to foreign competition to the extent that their prices are set abroad. Third, the Swedish producers are also assumed to be technology takers. Altogether, this implies that changes in comparative advantages can be expected to be met more or less only with quantity adjustments in the value added processes of given engineering products. Changes in comparative advantages are, presumably, exogenously given and caused by altering relative prices, differential market growth rates and technical change and differential (domestic) developments of factor supplies.

This set of factor proportions theories is analytically employed to seek for explanations of Sweden's engineering trade specialization and its long run development. Later on in the study complementary explanatory factors associable with certain arguments of the product cycle theory are also taken account of.

The pattern of trade specialization may be measured in many ways. Section 2.2 presents and defines the measures chosen for this study. The measures reflect the fact that supply-oriented causes are emphasized in the empirical analysis because of the focus on factor proportions accounts to the investigated specialization patterns. Unfortunately, the trade implications of the simple factor proportions theory with only two commodities, factors and countries are not easily generalizable. Section 2.3 seeks to clarify the consequences of this for the empirical analysis. Most factor proportions theories investigate the trade implications of differences between countries in factor abundance. The concept of factor abundance can, however, be defined in several ways. Section 2.4 discusses various factor-abundance definitions and describes the integration of this concept in the present study.

Likewise there are difficulties also with respect to the factor intensity concept. Section 2.5 provides the basis for the choice of interpretation. The empirical definition of this concept is given in chapter 4. The present chapter concludes with a section presenting the outline of the empirical analysis.

2.2 The trade specialization concept

Implicitly or explicitly, the measurement of trade specialization demands decisions on how to define products and to distinguish between exportables and importables. Many earlier studies (cf Leontief [1954, 1956], Keesing [1965, 1966, 1968a, 1971], Hufbauer [1970] avoided explicit consideration of those aspects by using a two-commodity framework. Recent studies of trade specialization have paid more attention to those issues (cf. Baldwin [1971], Branson [1972], Branson & Junz [1971], Carlsson & Ohlsson [1976], Harkness & Kyle [1975] and Ohlsson [1973]. The present study advances several definitions of a product by operating at different levels of aggregation and with two kinds of statistical classification systems. This can be seen as a precautionary measure to obtain a better control over the consequences of analysing heterogeneous groupings of products. The exact definitions are given in subsequent chapters.

Let us at this point take the product definitions for granted in order to see how the measurement of trade specialization can be made. Suppose first that there are n products for which no stocks are kept. Then the following identity holds :

$$O_k \equiv C_k + X_k - M_k; \quad k=1,\ldots,n. \qquad (1)$$

where O = domestic production
C = domestic consumption
X = exports
M = imports

The commodity specialization of a country in the n products can be measured by the equivalent measures $O_k - C_k$ and $X_k - M_k$. For homogeneous tradable (and traded) goods, $X_k = 0$ whenever $M_k > 0$ and $M_k = 0$ whenever $X_k > 0$. In practice, however, trade statistics consists of heterogeneous groups of commodities. If the degree of heterogeneity varies between commodity groups this variation will in itself influence the differences in measured specialization in a way which is not easily foreseen a priori. Heterogeneity differences are especially pronounced among the engineering industries. The fact that the impact of these differences on $X_k - M_k$ cannot be anticipated suggests the possibility of a distorted relationship between this measure of specialization and other possible, supply-oriented explanatory factors.

Non-tradable, homogeneous goods produce identical values for O_k and C_k, i.e. $X_k - M_k = 0$. The same value is, however, also obtained for heterogeneous commodity groups, with the same level of exports and imports. In practice, the tradability of commodities differ from low, but not zero, tradability to high, but not perfect tradability. Unless this cross-sectional variation in tradability can be measured and explicitly included as an explanatory factor, it may produce distorted relationships between $X_k - M_k$ and other explanatory factors. The variation in $X_k - M_k$ across commodity groups with net exports (or with net imports) is to a large extent explained by the very large variation in market size for these groups. Thus, even zero exports and production of, say, nails and bolts cannot produce net imports which are larger than net imports of, say, cars even when exports of cars are of almost the same magnitude as imports. This property of the net exports measure is theoretically unsatisfactory. Moreover, it is intuitively unfortunate, since such market size causes of trade specialization differences measured in this way are trivial. However, expression (1) can be easily adjusted to produce better specialization measures in this respect by simply dividing both sides by C_k;

which yields:

$$\frac{O_k}{C_k} \equiv 1 + \frac{X_k - M_k}{C_k} \qquad (2)$$

Expression 2 contains two equivalent trade specialization measures which are normalized for market size differences between commodity groups. Possible relationships between specialization measured in this way and market size can be given non-trivial interpretations based on existing trade theories. But do these measures also help to solve the deficiencies related to commodity group variations in tradability and heterogeneity? In order to answer the first part of this question some distinctions must be made between different causes of restricted tradability. A reasonable analytical simplification in the present study is that *country-specific* natural trade barriers which are connected with, for instance, Sweden's relatively remote position in Europe can be disregarded. The reason is that they tend to restrict imports and exports alike and for all types of products.

Commodity specific natural trade barriers like transportation costs are in principle more important since they impede trade differently for different commodities. From the point of view of measuring the trade specialization of a single country with one single measure they have, however, one fortunate property: both exports and imports of the product with the high natural trade barrier tend to be lower. In other words, these barriers tend to have a smaller impact on the cross-commodity group variation in the above (X-M)/C ratio than on X/C and M/C separately. However, since they reduce the volume of trade the absolute value of this net exports/consumption ratio tends to be lower for commodity groups with higher commodity-specific natural trade barriers.

Commodity-specific, policy imposed trade barriers in Sweden restrict (directly) only imports of the selected commodities whether they are tariff or non-tariff barriers. Only indirectly do they tend also to lower exports but then not necessarily more for exports of these selected commodities than for others. Therefore, such policy imposed barriers will usually distort the net exports/consumption ratio in a more serious way than natural ones from the point of view of finding supply-oriented explanations of the commodity trade specialization pattern.

The obvious remedy for all troubles related with tradability differences would be to include measures of such differences beside other explanatory variables in the analysis of specialization patterns. In practice, however, only nominal tariffs could be measured at the levels of aggregation investigated in this study. The remaining possibility was then to construct the specialization measures so as to be more insensitive to natural trade barriers and particularly commodity-specific ones. Since these barriers reduced the volume of trade relative to domestic consumption through both exports and imports this might simply be achieved by dividing expression (1) by $X_k + M_k$ instead of by C_k,

$$\frac{O_k - C_k}{X_k + M_k} \equiv \frac{X_k - M_k}{X_k + M_k} \qquad (3)$$

The term on the right hand side in this expression will henceforth be called the *net export ratio*. This specialization measure is at least to some extent normalizing for both non-measurable tradability differences and differences in market size. However, the measure is not as perfectly normalized for the latter differences as in the case of the ratio $(X_k-M_k)/C_k$. Beside empirical reasons for preferring $(X_k-M_k)/(X_k+M_k)$ to $(X_k-M_k)/C$, there is an additional theoretical reason behind this choice. The former ratio has the theoretically desirable property that pure exportables (respectively pure importables) obtain only one value on this ratio, namely +1,0 (-1,0).

In empirical analysis there are, however, no such homogeneous commodities. All industries or groups of commodities are more or less heterogeneous. It can be assumed that the larger the heterogeneity is in terms of trade determining factors, the smaller will the absolute value of the net export ratio tend to be. It is this property of the net export ratio, which explains its use for the measurement of the extent of so called intra-industry trade[1].

A high degree of heterogeneity of the trade determining factors implies, however, that measures of these factors also tend to cluster around their mean values. This is a desirable property for an investigation of possible determinants of trade specializations patterns, since the a priori unknown variations in heterogeneity will not tend to seriously distort the relationships between these patterns and their determinants. Throughout the study two-way trade is therefore treated as an aggregation problem. The use of the net export ratio in the analysis of the inter-industry pattern of specialization is reasonable provided that either one of two conditions is empirically fulfilled, namely :

1) the inter-industry variations of the ratio reflect differences in competitiveness substantially better than differences in heterogeneity

or

2) the inter-industry relationships between the net exports ratio and factor intensities do not depend on varying heterogeneity and possible intra-industry specialization *and* the Swedish industry factor intensities are representative of those of the rest of the world in spite of possible intra-industry specialization.

These points motivate the emphasis laid in this study on aggregation problems. However, in contrast to studies of intra-industry trade phenomena *per se* the methodology developed here is designed to help evaluate whether there are factor proportions explanations to the Swedish inter-industry trade patterns.

Although the net export ratio is accepted as our main measure of trade specialization, results are also presented for two additional measures with some theoretically desirable properties. Put $H_k = C_k - M_k$, i.e. that part of domestic production which is sold on the domestic market, and formula (1) reduces to :

$$O_k \equiv H_k + X_k \qquad (4)$$

That formula can be rewritten as follows :

$$O_k \equiv C_k \frac{H_k}{C_k} + C_{w,k} \frac{X_k}{C_{w,k}} \tag{5}$$

where H_k/C_k may be called the *home market share*, $X_k/C_{w,k}$ the *foreign market share* and $C_{w,k}$ = world consumption of product k (excluding that of Sweden). Recall that identical demand patterns of the two countries are assumed. Thus, since $C_{w,k}$ is a multiple, say α of C_k formula (5) reduces to :

$$\frac{O_k}{C_k} \equiv \frac{H_k}{C_k} + \alpha \frac{X_k}{C_{w,k}} \tag{6}$$

where α is a constant over all n goods.

It was argued above that O_k/C_k is sensitive to natural and policy imposed, commodity specific trade barriers. Note also that $X_k/C_{w,k}$ is actually 1 - $(H_{w,k}/C_{w,k})$, $H_{w,k}/C_{w,k}$ being the home market share of the rest of the world. Assuming no factor reversals and consistent factor abundance definitions, we would expect H_k/C_k and $X_k/C_{w,k}$ to vary similarly with the factor intensities if each product were perfectly tradable.

Natural and policy-imposed commodity-specific trade impediments will cause the two classes of relationships to differ from each other unless both types of impediments are uncorrelated with each factor intensity. The policy imposed barriers could be expected to vary inversely with the comparative advantage of each country and the relevant factor intensities[2]. Applied skillfully in both countries, they will tend to diminish systematically the incentives to trade, thus reducing exports from each country of its exportables. Still, we would normally expect the same sign of the two classes of relationships between the two trade specialization measures and the factor intensities.

In contrast, the existence of natural trade barriers for products with high (or low) values for one (or more) particular factor intensity can obviously produce different signs for the same specialization relationships. For heterogeneous products H_k/C_k will increase (decrease)[3] owing to decreased (increased) imports from the rest of the world. At the same time its exports and thus the ratio $X_k/C_{w,k}$ will decline (rise). For this reason, the existence of natural trade barriers reduces the otherwise presumably good correlation between H_k/C_k and $X_k/C_{w,k}$ and, even worse, may produce different signs when regressing the two ratios against the relevant factor intensity.

Obviously, it was impossible to estimate $C_{w,k}$. Instead $X_{w,k}$ = world exports of product k could be measured for one of the two classification systems used (by total OECD exports). Owing to the existence of natural and policy imposed trade barriers $X_{w,k}$ may not correlate particularly well with $C_{w,k}$ (or, for that matter, C_k). While the desirable tie to the theoretical assumption of identical demand pattern is lost for the measure $X_k/X_{w,k}$, the world export share, it may instead have the desirable property of being less sensitive to commodity-specific natural trade barriers on the part of the rest of the world[4]. However, the home market share *(H_k/C_k)* may still deviate by the same token from the *world export share ($X_k/X_{w,k}$)*. In conclusion, we consider the net export ratio *($[X_k-M_k]/[X_k+M_k]$)* and the world export share *($X_k/X_{w,k}$)* to be the two most reliable measures of specialization given the purpose of our study. Nevertheless, the home market share (H_k/C_k) will also be analyzed in order to preserve some flexibility and gain more information about the trade determinants. The net export ratio in domestic consumption and some other specialization measures are used only in chapter 7 which aims at revealing the adjustment process of new, exiting and permanent firms, respectively.

2.3 The rôle of the number 2

The empirical study of factor proportions explanations to the trade specialization patterns of a country recognizes explicitly more than two goods and factors but retains the two country assumption of the highly aggregated 2x2x2 factor proportions models. It is wellknown that the trade implications of the latter models are not generalizable to multicommodity, multifactor models. The following discussion aims at clarifying in what respects these implications can be carried over to such models and how remaining intricacies are empirically approached. First, however, the retention of the two country assumption is discussed from the point of view of its consequences for the factor abundance definition and for the associated possibilities of making certain simplifying small country assumptions.

The assumption of 2 countries. Allowing more than two countries would have affected the analysis in essentially three different ways. First of all, it might have meant a change in the definition of Sweden's factor abundance. As Bhagwati [1965] argued one would have to choose between comparing Sweden's relative factor endowment or prices with that of one of the following constellations of countries :

1. *The sum* of all other countries.
2. *The sum* of all countries with which Sweden had a direct exchange of goods.
3. *The sum* of all countries with which Sweden had a direct or indirect exchange of goods.
4. *Each one* of the above mentioned countries.

Choosing from among these and other possible comparisons of Sweden's factor abundance would have been extremely difficult owing to the scarcity of published theoretical and empirical multicountry studies with such alternative factor abundance definitions.

The second consequence of explicitly allowing more than two countries is also due to this scarcity; we do not know whether or not the inter-country trade flows can be associated with any of the possible definitions of a country's factor abundance.

The third consequence of bringing more than two countries into the analysis is related to the fact that Sweden is a small country compared to the rest of the world but not necessarily so if compared to subgroups of countries. A two country model thus motivates two very essential, simplifying small country assumptions while a multicountry model might have necessitated the abandonment of the same assumptions. The two assumptions are as follows.

First of all the small country assumption means that world market prices can be treated as exogenously given although not necessarily constant over time. Even a drastic change in Sweden's factor supplies would not imply any changes in commodity prices externally or internally (in free trade). Secondly, production technology and changes therein could at least as a first approximation be treated as exogenously given.

Consequently, the small country assumption preserves a closer relationship between the empirical analysis and already developed theories. In a multicountry framework it might have been more relevant in the study of some bilateral trade patterns to abandon the assumptions of exogenously given commodity prices and technologies. Even worse from the point of view of the underlying theoretical structure this might have been motivated only for certain subgroups of commodities[5].

From the point of view of finding powerful explanations the decision to rely on a two country framework and the related small country assumption may or may not be unfortunate. The results of the subsequent chapters provide information to a better evaluation of this[6].

The assumption of more than 2 goods. As this study analyses engineering trade specialization, there is no other option than the one involving more than two goods. It is well-known that for nx2x2 models the factor proportions theorem does not hold. However, it is still possible to rank the goods according to their comparative advantages by ranking of their factor intensities[7]. Three implications follow :

1. At least the good using most intensively the abundant (scarce) factor must be an exportable (importable).
2. Select arbitrarily a commodity and find out whether it is an exportable or an importable. If it is an exportable (importable) all other goods using more intensively (extensively) the abundant factor must also be exportables (importables). Alternatively :
3. Classify each commodity as an exportable or an importable. Each exportable must then use more intensively the abundant factor than each importable.

Apparently, the relaxation only of the assumption of two commodities does not mean that the factor proportions theory no longer generates clear-cut implications. Furthermore, nx2x2 models preserve in principle a simple definition of the factor abundance of the country. However, questions arise whether implications 2 and 3 are formally testable ones. Two different empirical observations suggest that they are not.

First the existence of both exports and imports of practically all items of trade nomenclature systems renders it difficult though not completely impossible to classify products as exportables and importables. The testability of both implication 2 and 3 above depends ultimately on whether or not reliable classifications can be obtained (see section 2.2 above for a further discussion and also Ohlsson [1975]).

Secondly, the implications do not hold in a situation of factor price equalization (see the discussion in Bhagwati [1972], Jones [1956-57], Melvin [1968],[1971a], Steward [1971] and Travis [1972]). According to the standard assumptions factor price equalization is implied. On the other hand such an equalization is obviously far from accomplished, which indicates that some assumptions are far from being good approximations of prevailing conditions. Since the assumptions differ much with respect to their consequences for the structure and implications of the theory it is not possible to judge on *a priori* grounds whether or not the problems connected with factor price equalization in theory and in practice are serious ones with respect to the above three implications.

In the empirical analysis the emphasis is on the rôle of trade impediments, the empirical definition of a "product", and factor intensity reversibility and its relationship with inter- and intra-industry specialization. A discussion of these points is included in the following sections.

The assumption of more than two factors. The explicit allowance of more than two factors has severe consequences for the concepts of factor abundance and factor intensities as well as for the trade implications[7].

In two-factor models the factor abundance of a country can be defined alternatively from its relative aggregate factor proportion or from its relative factor price ratio[8]. Recognizing more than two factors means that the comparisons between the two countries become more complicated. A country may be considered abundant in one particular factor if it is compared with another factor but not if compared with a third, which makes the use of factor endowment or price ratios dubious in defining a country's factor abundance.

On the other hand, no other device has been theoretically derived, and even if it had the striking scarcity of data would not have allowed us to use it. Accordingly, the only possible way out of the dilemma is to make use of as many denominators in the factor endowment (and price) ratios as the existing data allow in order to reach robust conclusions for at least a few of the factors.

Having more than two factors affects the factor intensity concept in a similar way. However, even if it turned out to be possible to arrive at a consistent strong hypothesis regarding the factor abundance of Sweden there would still remain problems regarding the measurement of factor requirements. This is due to the fact that more than two factors admit the existence of complementarity between factors[9]. The increased use of one production factor may always be associated with an increased use of another. For the empirical analysis factor complementarity may have two implications. The first one occurs if two factors are complementary in all sectors of production but not both either abundant or scarce. Then it will not be empirically possible to detect their opposite impact on the pattern of trade. Moreover, in the engineering sector the definition of a product in terms of activities may differ substantially between countries. In a small economy the domestic part of the production of a product may be more narrowly specialized in certain types of activities while the large economy have a more complete basket of activities. Accordingly if one of the two factors which are complementary in the integrated production line is scarce in Sweden and the other abundant, an intra-industry activity specialization can be the result. This may in turn in extreme cases lead to factor reversals at the industry or plant level. Note, however, that these reversals occur as a consequence of the mechanisms of the factor proportions theory. Only an incorrect application of the theory may in this case cause a (false) rejection of that theory. Apparently, the exact definition of the concept "product" is of importance for the outcome of the empirical analysis.

We shall return to this point in section 2.5 below.

The second empirical complication associated with factor complementarity occurs if two (or more) factors are complementary in some types of production while they are substitutes in others. If in addition one of the factors is abundant and the other scarce the consequence may be non-representative factor intensities in the country due to differences in intra-industry specialization. Empirically, consideration must in this case be paid to the construction of the measures of trade specialization (see further section 2.2 above and chapter 5).

To sum up, allowing for three or more factors motivates four precautionary measures to preserve a close connection between the empirical analysis and the properties of the factor proportions theory. First, attention must be paid to the definition of a product and the activities of production. Secondly and related to this problem, the existence of factor intensity reversals due to differences in intra-industry specialization should be carefully investigated. Thirdly, several factor intensity definitions should, if possible, be tried in the analysis of trade specialization determinants. Fourthly, special attention should be given to these problems also in defining measures of trade specialization.

Apparently, the consequences for the factor abundance and factor intensity concepts of more than two factors suffice to destroy the simplicity in the 2x2x2 model. The relationships between trade specialization on one hand and factor intensities and the country's factor abundance on the other are no longer clearcut. Unfortunately, there are additional complications.

Even in the nx2x2 model it was possible to derive three trade implications which resemble fairly closely the original factor proportions theorem. In addition the implication on the factor requirement of total exports and imports still hold. In nxmx2 models, it is true that *if* a factor could be consistently defined as an abundant factor, the factor requirement in aggregate exports and imports would reveal a larger relative use of that factor in exports than in imports. However, the three other implications do not hold.

Since the comparative advantage of a country in a multifactor model depends on several factor intensities, the product most (least) intensive in any single factor does not necessarily become an exportable (importable). Thus implication 1 of the nx2x2 model is no longer true. For the same reason an exportable is not necessarily characterized by a larger intensity in a single abundant factor than each of the importables. Implication 3 above must accordingly be rejected for multifactor models. In addition, every other good than that exportable using more intensively a single abundant factor may not always be an exportable, i.e. implication 2 above does not hold either.

However, it is possible to reformulate implications 2 and 3 in a meaningful way and still obtain implications which resemble reasonably well the content of the original factor proportions theorem. Those reformulations depend on whether or not it is possible to classify consistently certain factors as abundant and certain others as scarce as well as on whether the factor intensity concept is in fact retainable (of Chipman [1966] p.30). Given this the following implications should hold :

1. Choose an arbitrary product and find out whether it is an exportable or an importable. Suppose it is an exportable (importable). Use arbitrarily one of the factors which has in all comparisons been classified as a scarce factor as the denominator in the factor intensity. Then it should be true, *ceteris paribus*, that all products using more (less) intensively than this exportable (importable) at least one of the abundant factors are also exportables (importables).

2. Classify each product according to whether it is an exportable or an importable. Use analogously as in point 1 one of the scarce factors as a numeraire in constructing the factor intensities. Form subsets of products with the same value on all factor intensities except the intensity of one abundant factor. Within a given subset or products it should then be true that all products classified as exportables have higher intensities of the given abundant factor than all importables.

Again the classification of products into exportables and importables appears to be one of the most important obstacles to empirical application of those two implications (cf further section 2.2). In multi-product-multifactor models the structure of demand plays a more fundamental rôle in determining whether and to which extent a product becomes an importable or an exportable. Accordingly, the classification method must take account also of that impact.

2.4 The factor abundance definition

In the preceding section it was argued that the adherence to the two-country assumption preserves a clear definition of Sweden's factor abundance. It should be defined in comparison with the factor endowments and factor prices of the rest of the world[10]. Only shortages of data have prevented us from using this definition.

There are, however, three other assumptions behind the factor proportions theorem which if abandoned, affect the factor abundance definition, namely the assumptions of a) perfect internal factor mobility and perfect external factor immobility, b) that all products are tradables and c) identical demand patterns for the two countries.

The assumption of perfect internal factor mobility and perfect external factor immobility is essential for the type of adjustments that occur after a change in for instance tariff barriers. External factor immobility is necessary to hold internal factor endowments constant and thus to preserve the factor abundance of the country. Throughout the study it is assumed that external migration of factors is small enough to leave factor endowments relatively unchanged. However, in one chapter we shall partially abandon the assumption of perfect internal factor mobility by analysing certain entry and exit barriers (see chapter 7). The rest of the study assumes perfect internal factor mobility.

The existence of partial or perfect non-tradability of goods affects the factor abundance definition seriously. How the definition is affected depends on whether or not a) the NTG (non-tradable goods) and TG (tradable goods) sectors use only *common* (primary) factors, b) there are inter-industry linkages between the sectors, c) the demand conditions for NTG's and TG's are identical for the two countries and d) there are differences in technology for the countries as regards the NTG's.

The existence of NTG's prevents a theoretically well-established definition of Sweden's factor abundance. Moreover, available statistical data do not allow any sophisticated empirical treatment of that problem. In fact, as will be made evident in chapter 3 practically all available data refer to factors employed within the industrial sector.

The third obstacle in defining Sweden's factor abundance properly is the rôle of non-identical demand functions for the countries. Specifically such differences serve to make the factor abundance in terms of factor supplies different from that in terms of factor prices. Despite this Sweden's factor abundance had to be defined as if there was no such difference (cf chapter 3).

An additional problem associated with the demand conditions of Sweden and the rest of the world is caused by our explicit study of changes in trade specialization over time. Even with identical demand patterns, the combination of, say technical change or changing world market prices and non-homothetic consumer preferences may create differences between the endowment and price definitions of factor abundance. Again, for data reasons no account could be taken of that problem. However, in analysing changes in Sweden's engineering trade specialization it was possible to allow for differences in domestic demand conditions for the products, although not between Sweden and the world.

In conclusion, it must be recognized that the serious theoretical problems in defining a country's factor abundance can not be met with appropriate precautionary measures in the empirical definition of Sweden's factor abundance.

2.5 The factor intensity definition

As mentioned above the intensity concept is not strictly applicable in multifactor models. On the other hand such models do not provide alternatives to the use of factor intensities. In order to counteract the risk for false conclusions three different variables were tried in the denominator of these intensitites (cf chapter 4 and appendix A).

Apart from this problem there are definitional problems associated with a) finding sufficiently homogeneous factors, b) the existence of intermediate goods, which has a bearing on the specification of products in terms of activities and c) factor reversals and internationally given production functions. Since the selection of factors for the analysis boiled down to more or less a matter of data availability it will first be discussed in chapter 3.

The key issue associated with intermediate goods in an empirical analysis is whether the main purpose is to investigate net factor flows through trade or characteristics of the type of products the country has specialized in[11]. In the former case one is interested in both the direct and indirect factor content but instead ignorant about separating between different kinds of exportables and importables, respectively.

Studies of patterns of specialization attempt in contrast to separate products (=technically non-separable activities) or at least groupings of such products with similar technologies. Obviously, non-separable processes may in rare cases be the whole production within an industry, in others the combined production of several industries[12] and in yet others, parts of industries.

For specialization studies it is difficult to judge whether one should use only the direct use of the factors, total factor use or something in between. Several works have discussed the consequences for the Rybczynski theorem of intermediate goods and the use of gross and net factor intensities (see Batra [1973], chapters 7 and 8, Batra & Casas [1973], Chang & Mayer [1973], Kemp & Uekawa [1972] and Ray [1972]). Two conclusions can be drawn. First, it matters whether the intermediate goods are introduced as what is called "inter-industry flows" or as "pure intermediate products" (cf Batra [1973] chapters 7 and 8). In the latter case gross and net factor intensity rankings may diverge. Secondly, the trade implications are not affected as much in "inter-industry flows" models as in "pure intermediate goods" models. In the latter, the so called Rybczynski theorem may hold for gross factor intensities but not for net factor intensities (cf Batra [1973] chapter 8).

The present study is based on the inter-industry flows notion. There are three empirical reasons for this choice. First, the input-output relationship within the engineering sector look very much like inter-industry flows[13]. Secondly, apart from deliveries within that sector the main supplier of intermediate goods is the metal industry and in particular the iron and steel industry. Thirdly, practically each subindustry of the engineering and metal industries has relatively large export and/or import ratios and none is lacking completely in foreign trade. In this respect almost each industry appears to have separable activities from all other ones.

Nevertheless, it might be interesting to illustrate empirically whether factor intensities in the gross and net sense in fact deviate much from each other. Carlsson & Ohlsson [1976] have published a correlation matrix showing correlation between direct (*D*), i.e. net, factor intensities and total (*T*), i.e. gross, factor intensities according to the 1957 Swedish input-output matrix (Höglund & Werin [1964]) which is still the most comprehensive one. The following correlations were obtained between direct and total factor intensities in the manufacturing industry (106 sectors) :

Capital per employee	0.89
Engineers per employee	0.92
Blue collar workers per employee	0.86
Domestic forest raw materials per employee	0.76
Domestic iron raw materials per employee	0.24
Domestic non-ferrous ores per employee	0.27
Non-competing imports of raw materials per employee	0.97

There are strong relationships between direct and total factor intensities as regards the capital intensity and the two skilled labour intensities. In contrast, the intensities of domestic raw materials generate lower correlations especially the iron and non-ferrous metal intensities. The latter result is probably due to the fact that certain engineering industries use large amounts of metals indirectly but directly very little.

Carlsson & Ohlsson [1976] tested the above intensities and nominal or effective tariff rates in regressions against two types of specialization measures. One was constructed from *direct* exports and imports of the respective sectors (normalized by domestic production or consumption) and the other from total, i.e. direct and indirect exports of each sector (normalized by domestic production)[14].

It turned out that the coefficient for the capital intensity was not significantly positive if total or gross factor intensities were used in regressions against *direct* specialization measures. However, that appeared not to be attributable to differences in the ranking of gross and net capital intensities but rather to the fact that the gross capital and gross iron raw material intensities were highly correlated (r=0.54) whereas the corresponding net intensities were not.

Concludingly, the results seem to support the use of a theoretical framework for the engineering sector based on inter-industry flows between tradable goods sectors rather than pure intermediate models. At least there do not appear to be important differences in the ranking of the gross and net factor intensities relevant to each sector. Hence, the assumption of the standard 2x2x2 model with two final products produced in fully integrated production lines is abandoned in this monograph for the more realistic assumption of many separate production processes/products producing and using (in part at least) themselves as intermediate goods.

The early chapters of the study assume that the factor intensities are given internationally as well as intertemporally. This is in a two commodity model a very restrictive assumption. The occurrence of factor reversals is more likely in multiproduct-multifactor models than in 2x2 models[15], but on the other hand less damaging from the point of view of the trade implications.

The possible impact of factor intensity reversals motivates several analytical actions. One is to study whether heterogeneous industries combined with intra-industry specialization make Swedish industry factor intensities internationally non-representative (cf chapter 5). Another is to investigate the intertemporal stability of the Swedish factor intensities (chapter 6).

No effort is made to study the influence of non-constant returns to scale directly. However, some chapters use a methodology which may reveal some of the consequences of the combined rôle of economies of scale and a small domestic market (chapters 8 and 9). The assumption of an internationally known technology is accepted without any empirical analysis whatsoever[16].

2.6 Outline of the empirical methodology

The theoretical framework of the present study is based on multi-product-multifactor-two country models with inter-industry flows, tradable intermediate goods and with the country in question being comparatively small. Such models do not have clear-cut implications in the sense that the trade pattern in equilibrium depends only on the factor intensities of the products. However, the more abundant a given factor is the more likely it should be that products using that factor intensively are exportables. For this to be the case we have to find factors that are abundant according to all definitions of factor intensities, with denominators consisting only of truly scarce factors.

A critical part in investigating factor proportions explanations of a country's pattern of trade is therefore the empirical analysis of its factor abundance. Unfortunately, this analysis is also the most difficult one because of the lack of consistently defined international data on factor endowments and prices. Therefore it is often omitted in empirical work on trade patterns.

The present study uses a different approach. By combining results from earlier studies on factor endowments, factor prices and factor requirements in foreign trade a broad, but internally not very consistent basis for Sweden's possible factor abundance situation is obtained. However, these results cover a fairly long period. For this reason the factor abundance hypothesis rests on an assumption of a stable factor abundance during the period.

Obviously, a weak underpinning of Sweden's comparative advantage leaves much room for interpretation of the results of the analysis of factor proportions explanations to the engineering trade specialization in 1970. Since this analysis does not yield clear-cut results the empirical investigations might have proceeded along several routes. Arbitrarily, the following one was chosen. First, the assumption of a factor proportions account to Sweden's engineering specialization was throughout the study treated as a maintained hypothesis. Secondly, the factor abundance hypothesis was *temporarily* also treated as a maintained hypothesis.

However, beside a false factor abundance hypothesis weak results of our first approach might also be attributable to methodological deficiencies or to the pursuing of too simplistic a factor proportions approach. Step by step each one of these altogether three types of causes for weak results in our first approach is investigated. Needless to say, the coverage is far from complete and somewhat arbitrary. In many cases empirical considerations played a more important rôle than theoretical ones, since the theory as discussed above is far from clear in its trade implications as soon as the 2x2x2 model is abandoned.

Footnotes

1. See Grubel & Lloyd [1975] for a thorough treatment and references, but also Gray [1973] , Hufbauer & Chilas [1974] and Ohlsson [1974].

2. Given that the factor proportions theory is true.

3. Depending on whether that factor is an abundant (scarce) factor of the home country.

4. Since both the numerator and the denominator are reduced by such barriers.

5. Moreover, in multicountry models it is more reasonable to expect factor intensity reversals attributable to peculiarities in intra-industry specialization (cf chapter 5).

6. The final choice was made after an analysis for nine industries of the fabricated metal products industry, which accounts for about 25% of the total engineering industry (cf Ohlsson [1973], especially chapter 5). That study contained material showing for each industry the development of a) the import share of Swedish consump-

tion, b) shares of Swedish consumption and total imports by Japan and non-OECD, i.e. low wage countries and c) Sweden's shares of omports to three EFTA and three EEC markets, North-America and Japan. By and large, there was a stability in the comparative development os Swedish market shares over the nine industries for the nine markets studied. That finding suggested that at least the changes in Sweden's industry specialization might have been similar enough in the 1960's to allow a two-country framework and instead emphasize the analysis of the impact of differences in product or industry characteristics. The results of the subsequent analysis are reported below in chapters 6, 7 and 9.

In contrast to the findings for the Swedish fabricated metal product industry the results of Baldwin [1971], Bharadwaj [1962], Bharadwaj & Bhagwati [1967], Rosefielde [1973] , Roskamp & McMeekin [1968], Tatemoto & Ichimura [1959] and Wahl [1961] suggest that an analysis of trade specialization *in a given year* may benefit substantially from a study of bilateral trade flows. Their results can be compared with the results reported in chapters 4, 5 and 8 (and also Carlsson & Ohlsson [1967] as far as total Swedish trade in 1957 is concerned).

7. See Jones [1956-57] and Samuelson [1953-54] for the first generalization of the H-O-S model to multicommodity, multifactor models.

8. As mentioned above both alternatives should in principle refer to the autarky situation or at least the situation prevailing before a registered change in for instance factor supplies.

9. On this point we refer to Brown [1969], Hicks [1970], Nadiri [1970] and Sato & Koizumi [1973].

10. Of course, the mirror image of this advantage is the possible disadvantage of aggregation problems, especially since Sweden is not in an extreme position with respect to each one of the more than two factors.

11. Reference is here made to the discussion in Baldwin [1971].

12. Due to technical integration or input-output linkages between the industries combined with non-tradability of intermediary products.

13. It should be emphasized here that the definition of that sector excludes for instance repair shops, shipyards, and airplane industries. Also, the Swedish engineering industry produces relatively much of investment goods and less of consumer goods. Especially in the long run some

of the investments goods might in principle be looked upon as intermediate goods.

14. The analysis was carried out before the publication of Kemp & Uekawa [1972] which emphasized three production concepts; gross production, net production (value added) and a concept defined as gross production minus that part delivered to other sectors as intermediate goods.

15. Of Yeung & Tsang [1972] and also the discussion in Arrow, Chenery, Minhas & Solow [1961], Keesing [1971], Leontief [1964] and Minhas [1962].

16. After all world trade of engineering goods came from the industrial countries to 90% or more in the 1960's. The spread of knowledge should be fairly rapid between those countries for so heavily trade-exposed sectors.

Chapter 3

SWEDEN'S FACTOR ABUNDANCE AND COMPARATIVE ADVANTAGE

3.1 Introduction

This chapter implements empirically a hypothesis of Sweden's factor abundance. As pointed out earlier, there are serious problems with abandoning the assumptions of two countries, two homogeneous factors, perfect tradability and stable factor supplies, technology etc. Nevertheless, we shall use data on Sweden's factor endowment, factor prices and factor requirement of trade as though there were no such complications and no ambiguity of measurement. Accordingly, the hypothesis as to Sweden's abundant factors in the 1960's has a rather weak empirical foundation.

The selection of factors of production for the study utilized two simple criteria. Among the measurable factors those were chosen which a) varied much among engineering industries with respect to their share of production costs and/or b) according to earlier studies of Swedish factor supplies, factor prices or factor use in trade appeared to be abundantly or scarcely supplied. The exceptions to those rules were raw materials which were excluded for reasons explained in the preceding chapter. With these exceptions account should be taken to physical capital and skilled manual and non-manual workers. The following sections summarize previous findings on Sweden's endowment, prices and use in trade of those factors of production.

3.2 The Swedish endowment of human and non-human capital

There is only one international comparison on the relative endowment of non-human capital, namely Hufbauer [1970]. His estimates are for each country based on an accumulation of gross investment in the manufacturing industry 1953-1964, expressed in constant (1964) US dollars and divided by total employees of that industry[1].

According to Hufbauer's estimates Sweden was one of

the more capital abundant countries in Europe, surpassed only by Norway. Both Canada and the US turned out to have much higher capital proportions. Nevertheless, in a two-country framework which includes also non-industrial countries Sweden would probably have a higher capital proportion than the rest of the world.

The study of the endowment of human capital includes figures on the endowments of both separate skill categories and a combined skilled employee category. Data on the latter category are presented by Hufbauer [1970]. However, the Swedish figure is unfortunately not comparable to those of the other countries since it includes the relatively large employee group "foremen" [2].

Accordingly, Sweden was in all probability not the most richly endowed industrial country as far as skilled personnel is concerned in 1964.

Another source for evaluating Sweden's supply of human capital is OECD [1970] which contains statistics on both skilled labour categories and educational background of the labour force for ten OECD countries in the early 1960's. As regards the category "high level manpower" [3] measured in per cent of total labour force, Sweden had one of the highest percentages in Europe but not in the world. It ranked somewhat higher in terms of professional, technical and related workers (ISCO group 0) and still higher for scientific and technical personnel. The percentage of the labour force with university degrees was one of the lower among the ten countries and the corresponding percentage for scientific and technical degrees did not differ very much from other European countries. The fact that the Swedish labour force had a comparatively long average number of school years can be attributed to the fact that the least educated spend relatively many years in school[4].

The conclusion is that Sweden in the early 1960's was not one of the most abundantly endowed industrial countries with respect to highly educated, skilled non-manual workers. Nevertheless, a comparison between Sweden and the rest of the world would probably lead to the conclusion that it had a higher proportion of skilled personnel. A similar conclusion can be reached for the proportion of Swedish R & D personnel (or costs) for the years 1963, 1964 and 1971[5].

In conclusion Sweden was richly endowed neither with human nor non-human capital relative to total labour and compared to each one of the other major industrial countries in the early 1960's. However, it was probably one of the, say, five most human and non-human capital abundant countries. In addition, a comparison between Sweden and the rest of the world as a whole would probably rank Sweden as the more capital abundant area.

3.3 Swedish relative factor prices

Although there are indeed not many sources of data on factor prices, it is possible to gather some information as to Sweden's relative factor prices. Figure 3:1 shows total wage costs per hours worked for manual workers in manufacturing for selected industrial countries in relation to the corresponding Swedish figures. (Index = 100 for the Swedish total wage cost in each year.) According to figure 3:1 Swedish total wage costs for manual workers were one of the highest in the world in the 1960's, surpassed only by the US wage costs. Many European countries had wage costs which were 30-40 per cent lower than the Swedish ones throughout the 1960's.

Let us use total wage costs for manual workers as norm of comparison for comparing relative factor price ratios. To our knowledge there are no international comparisons of the price or cost of capital. It is sometimes argued that international differences in capital costs are relatively small due to the existence of international capital markets, international enterprises and trade in capital goods. If that is true, Sweden should in consequence have one of the highest wage/rent ratios in the world and in consequence a comparative cost advantage in capital intensive goods. Unfortunately, in the absence of empirical data on the price of capital this remains a speculation, widely believed by industry leaders.

Two sources of information on the relative salaries of Swedish technical personnel or engineers were available. According to Layton, Harlow & Hoghton [1972], the total salary costs for engineers or technicians in a multinational electronic company were about the same in Sweden, Italy and France in 1960. The corresponding levels in Great Britain and West Germany ranged at the same time between 58 and 75% respectively between 75 and 87% of the Swedish levels. A comparison with figure 3:1 suggests that the ratio salary of engineers/wages of manual workers was much lower in Sweden than in France and Italy and possibly also lower than in Britain and Germany.

Figure 3:1

Total wage costs per hour in the manufacturing industry in certain countries as percentages of the corresponding Swedish figures 1960-1970[a]

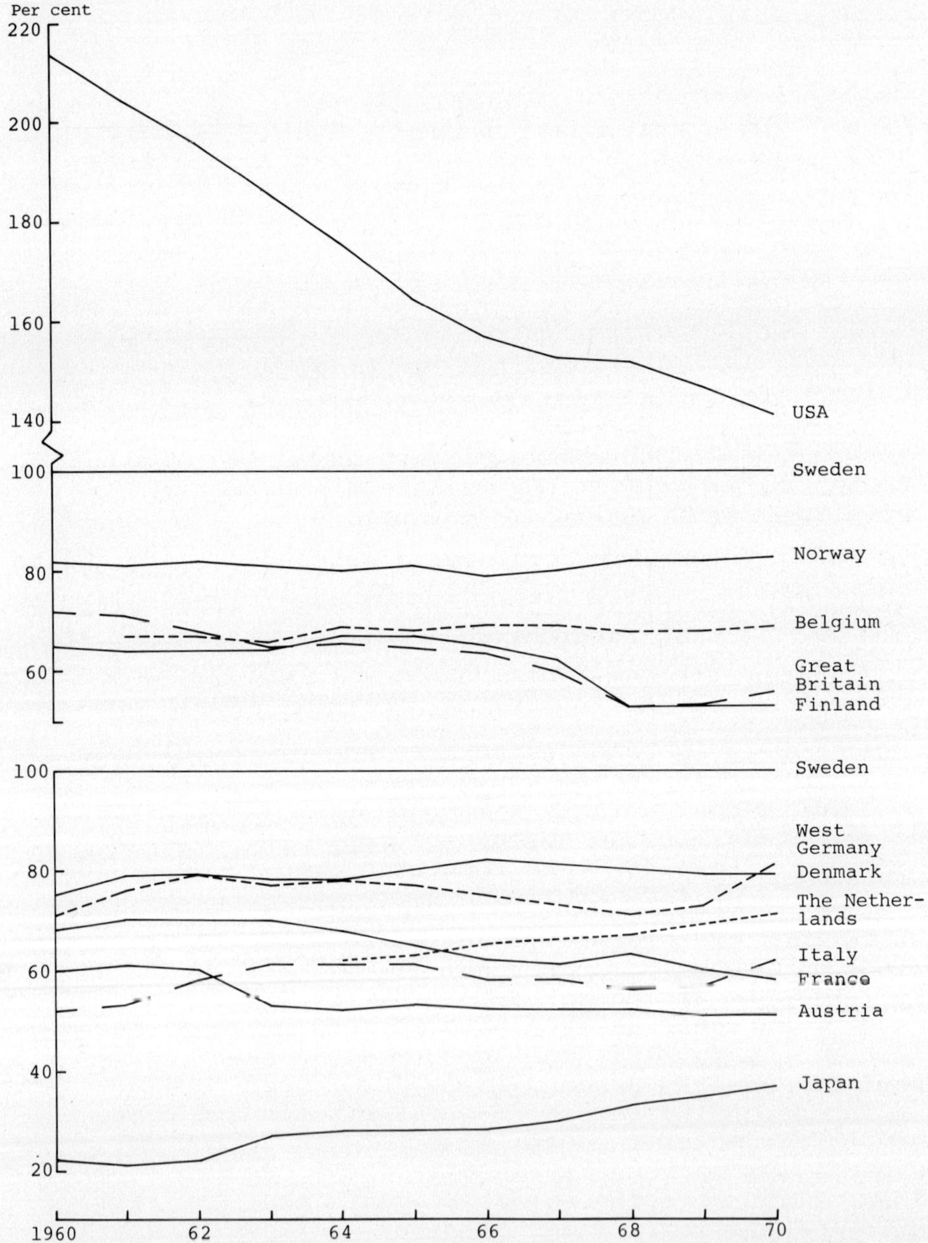

[a] Total wage costs include direct wage costs + indirect wage costs such as costs for vacation wages, social security, etc.

Source: SAF, Direct and Total Wage Cost for Workers - International Survey 1959-1970.

This comparison can be complemented with data from 1967 relating gross yearly salaries of so-called civil engineers (engineers holding a university degree or equivalent) to gross yearly wages of manual workers in the manufacturing industry (cf SACO [1971] table 4.2.1):

Sweden	Denmark	Finland	Norway	England	France	The Netherlands	West Germany	USA
402	396	455	294	246	503	429	275	265

According to these figures the relative salary of Swedish civil engineers is neither particularly high nor particularly low in a comparison with certain other industrial countries. If this conclusion holds also for other engineers as well as for technical personnel in general, it corresponds fairly well with Sweden's relative endowments of the same categories. Only if the comparison is extended to include all other countries as a whole will Sweden appear to have been abundant in technical personnel (or engineers) in the 1960's, both with respect to endowment and prices.

So far human capital has been discussed only for white collar workers. However, there are also large skill differences between categories of blue collar or manual workers. This was observed by Keesing [1968] who found the content of two categories of skilled manual workers in Swedish exports and imports to distinguish Sweden most from other countries. Fortunately, there are some data available on relative wages of skilled manual workers for Sweden. Ohlsson ([1973] figure 2:5) found a remarkable constancy over a fifty year long period in the ratio between wages for skilled manual workers and wages for unskilled manual workers within the Swedish engineering industry. Throughout business cycles, the world wide depression of the 1930's and World War II this ratio varied between 113 and 114%.

That difference is remarkably small in an international comparison. The Swedish Employers' Confederation collected statistical material on (gross) pay differences in 154 firms [6] between the simplest manual worker skills and the most complex ones (SAF [1973]; cf also Norstedt [1972]). According to that admittedly rough material, the complex skills group obtained a 30% higher pay in Sweden, a 60% higher pay in Western Europe and a 90% higher pay in the USA. Even though the underlying data base is a poor one, the finding that Sweden had very low comparative costs for skilled manual work-

3.4 The factor use in Swedish trade

Compared with data on Sweden's factor endowment and prices, data on its factor use in trade is fairly abundant. However, the available sources differ substantially with respect to methodology, measurement of variables and years covered. Estimates of the relative use of *non-human capital* in Swedish trade have been published by Carlsson & Ohlsson [1976] for the year 1957, Keesing [1965] also for 1957 and Hufbauer [1970] for 1965. Only Carlsson & Ohlsson use Swedish data obtained from the first and most comprehensive input-output study (cf Höglund & Werin [1964]). That study defines capital in terms of capital services measured for each sector as value added minus the total sum of wages and salaries. Defined in this way and using the Leontief method Swedish exports were found to have a 20% higher direct + indirect capital/employee than Swedish representative imports (unpublished results of Carlsson & Ohlsson [1976]). Both the export ratio of production and the net export ratio over domestic consumption were significantly positively correlated with the direct capital intensity of the sectors. However, as mentioned in chapter 2 the corresponding relationships with respect to total (direct and indirect) capital intensity were insignificant, possibly because of a strong correlation between this intensity and the intensities of domestic iron ore and domestic forest materials. The interpretation of Carlsson & Ohlsson was that capital was a complementary factor to those domestic raw materials in processes manufacturing those materials.

Keesing [1965] and Hufbauer [1970] follow in part the Leontief method in estimating the (direct) factor use in aggregate exports and imports, respectively. Using US capital requirements of production Keesing [1965] found Swedish exports in 1957 to have a somewhat larger content of capital per dollar value added than Swedish imports. Analogous estimates for eight other countries showed that only the USA and West Germany received larger relative differences between export and import capital contents.

Hufbauer [1970] found also but for the year 1965 a larger capital content in Swedish exports than in its imports. His study included 24 countries and Sweden's exports ranked as the sixth highest in terms of relative capital use while its imports ranked nineteen.

For a cross-section of 16 non-sheltered industries measures of Swedish specialization in 1965 were regressed on various factor intensities at an early stage of the analysis of the present study. Both the Swedish world export share and its export ratio of production were significantly, positively correlated with capital intensity measures [8].

The sources cited above give the overwhelming impression that Swedish exports are more capital intensive than Swedish imports at least in 1957 and 1965. This result does not seem to be sensitive either to capital measure or methodology used. However, there are some indications that the higher capital intensiveness in exports might be attributable to a specialization in manufacturing a few abundant domestic raw materials rather than to an across-the-board specialization in all capital intensive production sectors.

Several authors have studied the relative use of *human capital* in Swedish trade. Here we shall distinguish between the use of skilled manual production workers, skilled non-manual production workers and non-production workers used in R & D.

Keesing [1965] included estimates of the *skilled manual worker content* of manufacturing exports and imports in 1957 for nine countries. Sweden ranked as the second most intensive skilled manual worker exporter and the sixth most intensive importer. Five years later, according to Keesing [1968a], Sweden had developed as the most intensive skilled manual worker exporter among 14 countries. The latter study distinguished two categories of skilled manual workers, one of which was constructed to include modern skills in relatively common use while the other contained both more industry-specific modern skills and common but not modern skills. West Germany had a pattern of specialization similar to Sweden while the USA and Switzerland specialized in goods intensively utilizing only commonly used, modern manual skills.

The skilled, non-manual production workers content in trade was analysed by Carlsson & Ohlsson [1976] with respect to technical personnel (then including foremen). According to unpublished material of that study the share of such personnel in total employees was 5.2% in Swedish manufacturing exports and 6.8% in its manufacturing imports. Both the export ratio of production and the import share of domestic consumption were positively correlated with the (direct) technical personnel intensity. Consequently, the net exports ratio over domestic consumption was uncorrelated with that intensity[9] although a positive coefficient was obtained for the total (direct + indirect) technical personnel intensity.

Keesing [1965] found Sweden to have the highest export content of professional, technical and managerial skills in a comparison with eight industrial countries. Swedish imports ranked together with those of the Netherlands as number four with respect to this intensity. Keesing [1968a] repeated these calculations for the 1962 trade patterns of 12 industrial countries plus India but with four categories of skilled non-manual production workers: a) scientists and engineers, b) technicians and draftsmen, c) other professionals, and d) managers.

Swedish exports ranked in intensiveness as number 6, 5 6 and 8 respectively for these categories and its imports as 8, 6, 10 and 11. In contrast to 1957 Sweden did not in 1962 appear to be particularly specialized in skilled white collar personnel [10]. The latter conclusion holds also for scientific and technical personnel and engineers in Swedish exports of 1965 (cf Keesing [1971]).

Hufbauer [1970] utilized two measures of the human capital intensity— the number of scientists, engineers and technicians per employee and the average wage per year and employee. Both measures were derived from US statistics. Among 24 countries Sweden obtained for its 1965 exports the ranks 7 and 3 for respective intensity measures and ranks 16 and 15 for its imports. Fourteen of the twentyfour countries were industrialized countries. Hufbauer's results combined with those of Carlsson & Ohlsson [1976] and Keesing [1965,1968a] indicate that Sweden might be specialized in certain white collar skills in a global comparison but may not be so in a comparison among industrial countries.

Keesing [1968b] and Kenen [1970] emphasize the rôle of *R & D personnel* as a structural determinant of manufacturing trade. This skill category may not always be strongly tied to manufacturing processes. Their results indicated a strong influence on the US pattern of commodity trade specialization. Hence, it was worth examining whether the R & D intensiveness had a similarly important rôle for Sweden.

OECD [1970] contained two international comparisons of the export performance of certain industrial countries as regards R & D intensive compared to R & D extensive products. One comparison used a classification of industries while the other used a classification of commodity groups. Two groups of R & D intensive industries were distinguished: a) the drugs, chemical, electronical and instrument industries, and b) the same four industries plus the aircraft and non-electrical machinery industries.

The average Swedish shares of world exports in 1963-1965 of these two product (industry) groups were 2.2 and 2.8%, respectively, as compared with a total share for all goods of 3.5%. If instead the commodity group classification is used Sweden's world exports share of R & D intensive products was 4.0%. Apparently, the results are contradictory for Sweden.

In order to reach more definite conclusions a regression analysis of Sweden's specialization on 16 non-sheltered industries was carried out for the year 1965. Several measures of specialization and R & D intensities were used and the latter were included along with various combinations of other factor intensities. The only significant regression coefficient obtained for R & D personnel was a *negative* one between the export ratio of production and the proportion of development personnel in total employees. In conclusion, there are no indications that the R & D intensity has had the same fundamental rôle for the Swedish trade pattern as it appears to have had for the US one.

3.5 Conclusions about Sweden's factor abundance

According to results of earlier studies Sweden can be assumed to be abundant in certain raw materials (iron ore, forests and water falls) and in skilled manual workers. On the other hand Sweden appears to be scarcely supplied with low- and unskilled manual workers and other raw materials. The sources cited are not as unanimous as regards non-human capital and white-collar skills. However, Sweden appeared to be more abundant in the former than in the latter kinds of capital in comparison with other industrial countries. Given the two country framework used in this study it will be assumed that Sweden was abundant in both kinds of capital in the 1960's compared to the rest of the world.

Concludingly, our working hypothesis is that Sweden was abundant in skilled manual workers, skilled non-manual workers and physical capital[11]. This suggests that under certain assumptions Sweden had a comparative advantage and may have been specialized in products intensive in using those factors. The following chapter tests the latter part of the assumption on the inter-industry specialization pattern in 1970.

SNI no.	Industry	(1)	(2)	(3)	(4)
38291	Manufacture of household appliances except purely electric	24	70	41	79
382991	Lifting and hoisting machinery manufacturing[c]	23(28)	54(55)	57	119
38242	Construction and mining machinery manufacturing	23(27)	62(64)	49	98
38249	Manufacture of industrial machinery not elsewhere classified	23(19)	41(45)	70	153
38259	Manufacture of other office and accounting machinery	23	18	88	212
38195	Household metal ware manufacturing	17	73	34	65
38194	Manufacture of other metal products for construction purposes	17	75	32	60
382993	Manufacture of general purpose parts of machinery[c]	15	44	64	133
3813	Manufacture of structural metal products	12 (7)	94(93)	8	15
382992	Liquid pump manufacturing[c]	10	34	70	146
3821	Manufacture of engines and turbines	8 (8)	71(65)	32	63

Cont.

SNI no.	Industry	(1)	(2)	(3)	(4)
3822	Manufacture of agricultural machinery and equipment	8 (9)	70 (68)	33	65
3831	Manufacture of electrical industrial machinery and apparatus	6 (14)	58 (42)[d]	45	90
382999	Manufacture of other machinery and equipment[c]	5 (1)	56 (50)	46	92
3832	Manufacture of radio, television and communication equipment and apparatus[b]	3 (14)	48 (57)	53	107
38231	Metal-working machinery manufacturing	-2 (-8)	37 (34)	63	125
3833	Manufacture of electrical applicances and housewares	-7	71	26	54
38391	Manufacture of insulated wires and cables[b]	-9 (4)	84 (86)	13	28
38193	Nail, bolt and nut manufacturing	-14	74	21	45
38251	Manufacture of computing equipment	-14 (-22)	24 (27)	70	133
3851	Manufacture of professional and scientific, and measuring and controlling equipment, not elsewhere classified	-15 (-13)	30 (31)	63	121
38199	Manufacture of other metal products	-16	57	36	75

Cont.

SNI no.	Industry	(1)	(2)	(3)	(4)
38412	Boat building and repairing[a]	-19	66	26	57
38432	Manufacture of motor vehicle engines, parts and trailers	-22	57	33	71
38192	Wire cloth, wire and cable manufacturing	-24	83	11	27
3845	Manufacture and repair of aircraft[a]	-42(-10)	58(55)	22	60
38191	Manufacture of metal containers	-49	91	3	13
3844	Manufacture of motorcycles and bicycles	-56	49	22	65
38399	Manufacture of other electrical equipment	-65(-62)	40(45)	24	72
38393	Light bulb and fluorescent lamp manufacturing[a]	-83	40	12	66

[a] These industries are excluded from the following regression analysis for reasons explained in chapter 1.

[b] The average values for 1970 and 1971 are for the respective columns:

38241	46	30	86	260
3832	9	53	52	104
38391	-3	85	14	29
38413	53	77	50	100

Those values are used in the following regression analysis.

[c] Regressions including the skilled manual worker intensity must be run at the 5-digit industry level. Thus, these four 6-digit SNI-numbers are combined into the 5-digit no. 38299, for which the net export ratio and the home market share in 1970 were 12 and 52% respectively.

[d] This low value for 1971 is attributable to the lack of published gross output values for a number of important commodity items. Therefore the 1970 values are used instead of average values for the two years.

Note: The net export ratio is defined as the difference between exports and imports divided by the sum of exports and imports. The home market share is defined as production minus exports divided by domestic consumption. The export ratio is measured as the export share of production.

Sources: SOS, Industri, del 2 1970, 1971 and SOS, Utrikeshandel, del 2 1970, 1971.

According to the net export ratio Sweden was in 1970 generally specialized in transport equipment (excluding aircraft, bicycles and motor-cycles) and machinery equipment for industries manufacturing domestically abundant raw materials. Among industries with high net export ratios are the cutlery, hand tools and hardware industry, and the storage battery and accumulator industry. In the former industry certain Swedish steel and hard metal producers have obtained a strong foothold and the latter was established at the turn of the last century on the basis of an early Swedish innovation.

Sweden was in contrast not in general specialized in industries producing fabricated metal or electrotechnical products. Neither had it in 1970 concentrated on aircraft, bicycles and motorcycles or motor vehicle engines parts and trailers.

A comparison between the net export ratio and the home market share showed that they were not at all systematically related. The home market share was, however, negatively correlated with both the export ratio of production and the gross trade ratio over domestic consumption while the latter two ratios varied positively with each other. Such a pattern would be obtained if each measure depended strongly on tradability and if tradability differed substantially from one sector to another.

Suppose that the gross trade ratio could be explained largely by tradability. This would imply that the manufacture of structural metal parts, of metal containers, of wire cloth, wire and cable, and of insulated wires and cables would exhibit the lowest tradabilities among the engineering industries. According to an English input-output study for the year 1963 the share of transport costs in value added for counterparts to these four industries varied between 6 and 15% (cf

Edwards [1970], table 5 and appendix table). The corresponding percentages for fabricated metal products, machinery, etc., ships and other transport equipment were respectively 5.6, 1.4, 0.5 and 1.0%. Hence, it appears that a comparatively low trade exposure corresponds to comparatively high transport costs.

An even stronger case on this point could be made if it were possible to show that industries with large trade exposure also have relatively low transport costs in per cent of value added. Table 4:1 shows large gross trade ratios for industries producing pulp and paper machinery, ships and other office machinery. The latter two industries have relatively low transport cost percentages (cf Edwards [1970]). However, no such data can be obtained for the pulp and paper machinery industry. According to other information the intra-industry specialization is very strong for the pulp and paper industry as is indicated by its very low home market share and high export ratio of production.

In summary, there are indications that inter-industry differences in tradability, due at least in part to transportation costs, affect the home market share, the export ratio of production and the gross trade ratio. Hence, it seems as if the most reliable measure of international competitiveness, aside from tradability, would be the net exports ratio and not the home market share. Unless we are successful in finding independent variables to reflect tradability of products, our efforts are better directed towards explaining differences in the net export ratio than in the home market share.

4.3 The independent variables

Reasons have been presented for studying whether the factor requirements of the industries are influential determinants of the inter-industry pattern of specialization. Chapter 2 as well as the results of the preceding section motivate the inclusion of variables to measure the tradability of products. In the following, two groups of independent variables are discussed and empirically defined.

As mentioned in chapter 2 the use of a multifactoral framework means that there is no theoretical basis for measuring inter-industry differences in factor requirements. Even the factor intensity concept became less relevant. Three kinds of definitions of factor requirements have been tried. They differ from each other only with respect to the denominator. The denominators tried were a) total employees, b) low- and unskilled labour[6] and c) value added. In the main text we shall restrict the presentation to the first factor intensity concept. Certain results for the other two concepts are given in appendix A.

According to chapter 3 at least three kinds of numerators in the factor intensities should be tried, namely measures of non-human capital, skilled non-manual production workers and skilled manual production workers. Swedish industrial statistics provide three alternative measures of *non-human capital*. One is gross capital income defined as value added minus total wage sum. For reasons made evident by the trade adjustment mechanisms set out in chapters 6 and 7 that measure was considered inferior to the other two in an inter-industry analysis.

Non-human capital might then alternatively be represented by either the capacity of motive power (in horse-power) or electricity consumption which indicates the utilization of that capacity. ross-sectional comparisons of capital intensities based on these two indicators for a five-year period suggested that either one could be used. Omitting shipyards, inter-industry differences were very similar and no measure was superior to the other on all criteria used in the analysis (cf appendix A). Arbitrarily, the capacity of motive power was selected for the analysis. The inter-industry differences in number of horsepower per employee are shown in table 4:2.

Although the largest measured capital intensities are ten times the lowest ones, the largest ones do not even attain the average capital intensity of the whole manufacturing industry (=13.1 horsepower/employee). Consequently, the regression results for the capital intensity obtained for the engineering sector may not be representative of the pattern for all manufactured goods. Given the hypothesis of a capital abundant economy, there is a danger that the relative cost of capital in the most capital intensive engineering industries is too small to create a sufficiently strong comparative advantage.

Swedish industrial statistics did not leave much choice as regards indicators of *skilled non-manual worker intensiveness*. The proportion of managers was on average small and moreover did not vary much. The latter was true also for foremen and sales personnel. Preliminary results for these categories did not generate any significant coefficients, which explain why they are excluded in the following presentation.

Table 4:2

Factor intensities in 39 engineering industries in 1970

Ranks within parenthesis

SNI no.	Industry	Capital intensity in horse-power/employee (1)	Technical intensity in per cent of total employees (2)	Skilled manual worker intensity in per cent of hours worked by manual workers (3)	A measure of the relative skilled manual worker quality (4)
3811	Manufacture of cutlery, hand tools and general hardware	5.5(18)	5.1(33)	30(17)	1.15(16)
3813	Manufacture of structural metal products	4.4(23)	5.2(32)	56 (6)	1.14(19)
38191	Manufacture of metal containers	4.6(22)	6.0(31)	20(24)	1.17(12)
38192	Wire cloth, wire and cable manufacturing	12.3 (3)	6.5(29)	21(23)	1.13(22)
38193	Nail, bolt and nut manufacturing	7.0(10)	4.4(35)	20(25)	1.11(27)
38194	Manufacture of other metal products for construction purposes	7.3 (7)	5.0(34)	18(27)	1.15(18)
38195	Household metal ware manufacturing	4.2(25)	4.3(37)	10(33)	1.17 (8)
38199	Manufacture of other metal products	5.1(20)	4.4(35)	28(18)	1.17 (8)

Cont.

SNI no.	Industry	(1)	(2)	(3)	(4)
3821	Manufacture of engines and turbines[a]	(9)	(3)	(8)	(5)
3822	Manufacture of agricultural machinery and equipment	5.9(16)	9.4(24)	24(20)	1.11(27)
38231	Metal-working machinery manufacturing	6.3(13)	10.4(17)	62 (4)	1.15(15)
38232	Wood-working machinery manufacturing	6.4(12)	9.8(21)	48(10)	1.12(25)
38241	Pulp and paper mill machinery manufacturing	7.2 (8)	16.3 (6)	68 (3)	1.16(14)
38242	Construction and mining machinery manufacturing	6.6(11)	11.5(14)	53 (7)	1.13(21)
38249	Manufacture of industrial machinery not elsewhere classified	5.2(19)	10.2(18)	56 (5)	1.18 (4)
38251	Manufacture of computing equipment	1.7(36)	30.9 (1)	46(11)	1.08(34)
38259	Manufacture of other office and accounting machinery	1.6(38)	12.2(13)	22(21)	1.11(29)
38291	Manufacture of household appliances, except purely electrical	4.7(21)	9.7(22)	11(32)	1.09(33)

Cont.

SNI no.	Industry	(1)	(2)	(3)	(4)
382991	Lifting and hoisting machinery manufacturing	5.7(17)	9.0(25)		
382992	Liquid pump manufacturing	4.1(27)	10.0(19)		
382993	Manufacture of general purpose parts of machinery	9.3 (5)	15.0 (8)	40(13)	1.12(26)
382999	Manufacture of other machinery and equipment	9.5 (4)	14.0(10)		
3831	Manufacture of electrical industrial machinery and apparatus	6.0(15)	20.1 (4)	31(16)	1.16(13)
3832	Manufacture of radio, television and communication equipment and apparatus	1.2(39)	19.2 (5)	19(26)	1.18 (6)
3833	Manufacture of electrical appliances and housewares	3.8(28)	8.6(26)	18(28)	1.19 (3)
38391	Manufacture of insulated wires and cables	12.5 (2)	10.5(16)	10(34)	1.17 (8)
38392	Storage battery and accumulator manufacturing	3.7(30)	10.8(15)	15(31)	1.10(32)
38393	Light bulb and fluorescent lamp manufacturing	2.1(35)	6.3(30)	–[b]	–[b]
38399	Manufacture of other electrical equipment	2.3(34)	9.9(20)	15(30)	1.18 (6)

Cont.

SNI no.	Industry	(1)	(2)	(3)	(4)
38411	Ship building and repairing	12.7 (1)	12.9(11)	52 (9)	1.10(31)
38412	Boat building and repairing	3.4(32)	1.6(39)	36(14)	1.05(35)
38413	Marine engine manufacturing	3.5(31)	9.5(23)	32(15)	1.14(20)
38421	Railroad equipment manufacturing	7.4 (6)	12.6(12)	70 (2)	1.20 (1)
38431	Motor vehicle and chassi manufacturing	4.4(23)	14.7 (9)	16(29)	1.10(30)
38432	Manufacture of motor vehicle engines, parts and trailers	6.3(13)	6.8(28)	25(19)	1.13(22)
3844	Manufacture of motorcycles and bicycles	3.8(28)	7.0(27)	7(35)	1.13(24)
3845	Manufacture and repair of aircraft	2.4(33)	28.4 (2)	70 (1)	1.15(16)
3849	Manufacture of transport equipment not elsewhere classified	4.2(25)	2.0(38)	22(22)	1.20 (1)
3851	Manufacture of professional and scientific, and measuring and controlling equipment not elsewhere classified	1.7(36)	15.9 (7)	43(12)	1.17 (8)
Total engineering industry		5.7	11.7	34[c]	

[a] Only ranks are published to preserve secrecy for certain large producers.

Cont.

[b] Light, bulb and fluorescent lamp manufacturing firms are not included as members of the Swedish Engineering Employers' Association (Verkstadsföreningen).

[c] According to estimates in *SOU* 1973:29, table 10:6.

Note: Rank 1 is assigned to the industry with the highest value of the respective intensities. Measures of capital, technical personnel and total employees have been obtained from the *SOS*, Industri 1970 del I, and certain unpublished figures from Statistiska centralbyrån (the Swedish Central Bureau of Statistics). The last two columns use estimates of hours worked by skilled and unskilled manual workers as well as wage rates from a statistical source collected jointly by the Swedish Engineering Employers' Association and the Swedish Metal Workers' Union (Metallindustriarbetareförbundet). The exact definition of each intensity is given in the main text and in appendix A.

The key factor proportion as regards non-manual human skills appeared to be the number of technicians[7]/employee. This is not surprising since this intensity varies from a few per cent up to more than 30% of total employees. Moreover, technical personnel receive relatively high salaries compared to the average for all employees which means that their share in production costs varies substantially among engineering products. Differences in technical personnel intensities among engineering industries cover fairly well the differences in the same intensity for the whole manufacturing industry. The average intensity for the engineering sector was 12% which compares with 8% for the manufacturing industry.

It must be emphasized that technical personnel in Swedish industrial statistics are personnel working at the plant. Engineers or technical personnel utilized for sales activities, R & D work etc. are not always or even in general included in this category.

The last category of production factors is *skilled manual workers*, which in chapter 3 appeared to be a key factor for the Swedish trade pattern. It can be expected to be an even more strategic factor in the engineering industry given the large variation in the proportion of skilled manual workers. However, Swedish industrial statistics offered only the wage per hour for manual workers as a potential proxy for that proportion. This variable was first used but the difficulties inherent in obtaining an unambigous interpretation of the results led us to abandon it[8].

In order to avoid those difficulties another statistical data base was utilized to construct measures of the skilled manual worker intensity. This data base is presented in appendix A. In short, the data base includes statistics from engineering plants for individual manual workers classified with respect both to types of occupation and to whether they are skilled or not. A worker is regarded as a skilled worker if a) he/she has a three year schooling in a particular skill and b) if the job he/she holds also requires that schooling. These quarterly statistics were connected with the new ISIC-based Swedish industrial classification system for production plants starting only in the 4th quarter of 1972 and then only down to the 5-digit level. The plant data were aggregated to the 5-digit industry level for the present study.

Two different skilled manual worker concepts were tried, one being all skilled manual workers and the other skilled manual workers in so-called core skills for the engineering industry. Both were related to total hours worked by all manual workers in the 4th quarter 1972. The correlation coefficient between the two skilled manual worker intensities was as high as 0.997. In the following, the skilled manual worker intensity used is that which includes all occupations.

According to SOU 1973:29, table 10:6, the average skilled manual worker intensity was about 34% in the engineering industry as a whole. That value is judged to be substantially higher than a correspondingly measured intensity for the whole manufacturing industry.

Table 4:2 shows that the skilled manual worker intensity ranges from 7% up to 70% of total hours worked by manual workers. Evidently, the skilled manual worker share of total production costs varies a great deal between different kinds of engineering goods.

However, skilled manual workers are far from a homogeneous category as regards the skill level. As a complement to the intensity measure,we choose the ratio between the hourly wage for skilled manual workers and that of unskilled manual workers[9]. Table 4:2 shows that the inter-industry differences are fairly small.

As discussed in chapter 2 explicit acknowledgement of more than two factors means that we may have complementary factors which do not vary independently. The problem would be especially serious if two factors tended to be used in a way so as to produce systematic covariation between the respective factor intensities over all industries[10]. The following correlation matrix for 31 engineering industries of table 4:2 shows whether multicollinearity is a potentially serious problem[11].

	(1)	(2)	(3)	(4)
(1) Capital intensity	1.00			
(2) Technical personnel intencity	-0.22	1.00		
(3) Skilled manual worker intensity	0.10	0.36	1.00	
(4) Percentage of small sized plants[12]	0.03	-0.63	0.15	1.00

There is a weak negative relationship between the capital and technical personnel intensities. Apart from certain industries in the elctrotechnical industry (SNI 383) the technical personnel and skilled manual worker intensities vary positively with each other. The correlation coefficients are not so large as to endanger the regression coefficients because of multicollinearity. As shown in the above matrix technical personnel are most intensively used in industries with a small proportion of small plants, i.e. plants with at most 200 manual workers[12]. Consequently, the technical personnel intensity also reflects inter-industry differences in plant size, the rôle of which is further analysed in chapter 7.

It has been emphasized above that differences in *tradability* appear to have substantial influence on certain measures of inter-industry specialization. Unfortunately, the only measurable determinant of tradability is Swedish import tariffs in 1970. Baldwin [1971] and Lundberg [1976] have amply demonstrated the difficulties in estimating non-tariff barriers to trade. No data were available on the inter-industry variation in transport costs.

Lacking input-output data for the industries it was impossible to calculate effective tariffs. However, in a general equilibrium analysis it is doubtful whether effective tariffs should be preferred to nominal tariffs. Earlier empirical studies have in any case for Sweden shown a close correspondence between the inter-sectoral structure of these tariff measures[13] (cf Carlsson & Ohlsson [1976] and Lundberg [1976]). The average (nominal) tariff level for imports was computed as total tariff revenues divided by total imports. This average tariff level is a weighted average of tariff rates for individual goods and countries provided that the latter rates are neither prohibitive nor redundant[14]. According to appendix table B:2 the level of the 1970 industry tariffs do not give reason to believe that the tariff rates are prohibitive. The average tariff level (weighted) for the engineering industry as a whole was only 5% with only one industry tariff substantially exceeding the 6% level, namely the motor vehicle industry (13.3%).

The structure of tariffs may indicate which industries were once regarded as non-competitive ones. Earlier studies have shown Swedish pre-EFTA import tariffs to be higher, the higher the labour intensity of the sector (cf Carlsson & Ohlsson [1976] and Lundberg [1976]). The engineering tariffs in 1970 were only weakly correlated with the capital intensity (-0.20)[15]. The correlation coefficients with the technical personnel and skilled manual worker intensities were also small (respectively 0.29[15] and -0.24). A comparison between the tariffs of 1960 and the capital intensity of 1970 generates a significant negative correlation coefficient (-0.35). Accordingly, there are indications of a tariff protection before the EFTA-period even within the engineering sector that in itself supports the hypothesis of a capital abundant Swedish economy at that time.

The corresponding correlation coefficient with the 1970 technical personnel intensity was 0.38. The analogous interpretation for that intensity would be that Sweden was not abundantly supplied with technical personnel before the 1960's. In both cases it is assumed that the factor intensities of 1970 and 1960 are irreversible (cf the analysis of that assumption in chapter 6).

4.4 Possible determinants of the inter-industry specialization patterns

4.4.1 Hypotheses

At the conclusion of chapter 3 it was assumed that Sweden was abundantly supplied with and had a comparative advantage in products intensive in a) capital, b) skilled non-manual labour, and c) skilled manual labour. In the following regression analysis we shall investigate whether the factor proportions theorem is generalizable to the multiproduct engineering trade pattern with several factor intensities. Our null hypothesis is that the net export ratio (and the home market share) is non-positively related to the capital, technical personnel and skilled manual worker (two measures) intensities.

Furthermore, the structural importance of Swedish tariffs is investigated. According to tariff theory the home market share should be higher, the higher the level of the tariff, provided that other variables reflecting differences in supply and demand elasticities are properly included. The former elasticities might be but not the latter since the only other independent variables of the regression analysis are the factor intensities. Nevertheless, our null-hypothesis is that of a non-positive relationship between home market shares and tariffs.

It was not possible to obtain a similar theoretical basis for including tariffs as an independent variable in regressions for the net export ratio. However, for historical reasons tariffs may have been structured in such a way as to maintain higher tariffs for the less competitive domestic producers (the diversification of production argument for tariffs). Hence, indirectly the net export ratio may be negatively correlated with tariffs before the 1960's. The subsequent tariff decreases towards EFTA countries and within the Kennedy round may have obscured that relationship, which is our reason for abstaining from any hypothesis regarding its sign in 1970.

4.4.2 Regression results

Table 4:3 presents the results of the regression analysis. The outcome for the net export ratio and for the home market share regressions differ substantially both with respect to explanatory value and the signs of the regression coefficients for the factor intensities. The factor intensities do not explain much of the variance of the net export ratio but do explain a substantial part of the variance of the home market share.

The net export ratio was not correlated with the capital and technical personnel intensities in 1970. It was positively correlated with the skilled manual worker intensity although that relationship did not manage to contribute much to the explanatory power. Only to the extent that Sweden's abundance is stronger in skilled manual workers than in capital or technical personnel are those results in accordance with our underlying hypothesis. With the net export ratio as our main dependent variable we must at this point conclude that the generalization of the factor proportions

Table 4:3

Regressions on the net export ratio and the home market share of 1970

Regression no.	Dependent variable	Constant	Regression coefficients (with standard error) for: K/L	L_T/L	L_Y/L_A	$L_{Y,Q}$	T	R^2	F-value (degrees of freedom)
1	$\frac{X-M}{X+M}$	0.021	0.008 (0.019)					0.005	0.174 (1;32)
2	$\frac{X-M}{X+M}$	0.016		0.430 (0.812)				0.009	0.281 (1;32)
3	$\frac{X-M}{X+M}$	-0.127			0.579^c (0.276)			0.132	4.413^b (1;29)
4	$\frac{X-M}{X+M}$	-0.134	0.003 (0.021)	-0.129 (0.923)	0.590^b (0.312)			0.134	0.392 (3;27)
5	$\frac{X-M}{X+M}$	1.269	0.006 (0.020)	-0.355 (0.968)	0.962^c (0.408)	-1.464 (1.718)	3.249 (2.769)	0.228	1.478 (5;25)
6	H/C	0.437	0.029^c (0.012)					0.154	5.843^c (1;32)
7	H/C	0.756		-1.426^c (0.507)				0.198	7.915^c (1;32)
8	H/C	0.697			-0.273^a (0.192)			0.065	2.021 (1;29)
9	H/C	-0.114	-0.028^c (0.012)	-0.819^a (0.553)	-0.409^b (0.233)	0.782 (0.982)	-2.053 (1.582)	0.443	3.970^c (5;25)

a = significant at the 10% level; *b* = significant at the 5% level; *c* = significant at the 2.5% level. $(X-M)/(X+M)$ = net export ratio 1970; H/C = home market share 1970; K/L = capital intensity 1970; L_T/L = technical personnel intensity 1970; L_Y/L_A = skilled manual worker intensity, 4th quarter of 1972; $L_{Y,Q}$ = the relative skilled manual worker quality, 4th quarter of 1972; t = average tariff on total imports 1970.

Note: None of the above variables is measured in per cent. The number of industries is only 31 in regressions including the skilled manual worker intensities for reasons discussed in tables 4:1 and 4:2.

theorem does not receive much support. It is not yet possible to determine whether this negative result can be attributed to an insufficiently strong abundance in capital and technical personnel or to the failure of traditional factor proportions theories to include the most powerful trade determinants. It must again be emphasized that what we have called the analogue of the factor proportions theorem in the 2x2x2 model is only one of many ways in which a country can exhaust its supply of abundant factors and ease the scarcity of others. Hence, it is too early to discard the factor proportions theory as our analytical framework.

As was noticed above the three factor intensities were much more powerful in explaning the Swedish home market share. That share tended to be larger the higher the capital intensity and the lower the technical personnel and skilled manual worker intensities. Again we have only one result which would lead to a rejection of the null hypothesis of non-positive regression coefficients, namely for the capital intensity. As regards the other two intensities the results instead indicate a scarcity of technical personnel and skilled manual workers in Sweden. Hence, regressions on both specialization measures lead to the same result for only the technical personnel intensity, for which the null hypothesis cannot be rejected.

A decisive conclusion can be reached for the capital and skilled manual worker intensities only if we have reason to believe more in one of the specialization measures. It has been indicated earlier that the net export ratio is the more reliable one on the ground that the home market share is more sensitive to differences in tradability. However, regression 9 gives evidence that the home market share is not significantly correlated with the tariff structure. If anything, that share tends to be larger the lower the domestic tariff level is!

A downgrading of the regression results for the home market share must in consequence rely on evidence of the importance of non-tariff barriers, either policy-imposed or natural ones. However, such a downgrading means either that we reject the analogue of the factor proportions theorem for the engineering trade pattern of Sweden or question the original factor abundance assumptions as regards the factors capital and technical personnel.

4.4.3 The possible rôle of tradability differences

It must be recalled that the information on non-tariff import barriers is scattered and of a qualitative kind. Here, we shall restrict ourselves to the rôle of the so-called S-labelling of electrical products and transport costs. The former barrier has been discussed by GATT and Lundberg [1976] in a way suggesting that it might be the most important policy-imposed barrier within the engineering sector.

The analysis of the two above mentioned barriers utilizes figures 4:1 and 4:2. Figure 4:1 shows for each industry the actual and estimated (from regression 5) net export ratios. Figure 4:2 shows the actual and estimated (from regression 9) home market shares. Let us begin with the S-labelling, which is a means to uphold very high safety standards on all electrical equipment sold and used in Sweden. The lack of an international safety standard and the small Swedish market have been said to constitute a non-tariff barrier to imports. Accordingly, the home market share should be high for almost all electrotechnical industries regardless of technology characteristics and comparative advantage. The net export ratio should better reflect differences in comparative advantage although it should also be affected positively by the limited imports. The export ratio of production is not necessarily affected.

According to figure 4:2 (and table 4:1) the home market shares are large for the insulated wire and cable industry, the storage battery and accumulator industry and perhaps also the electrical appliances and homeware industry. However, other electrotechnical industries do not have markedly high home market shares. Moreover, the net export ratio is positive for only one of the three above mentioned industries (the storage battery and accumulator industry) and the export ratio very low for all three. The products of all three industries appear to have a low tradability in both directions in spite of the differences in two of the factor intensities (cf table 4:2). They are all extremely extensive in skilled manual workers which contributes to a negative relationship between the home market share and the skilled manual worker intensity. Our conclusion is that the effects of the S-labelling cannot possibly distort the regression results for the home market share so drastically that we have to reject the regression results for that specialization variable completely. If anything it

Figure 4:1

Actual and estimated (from regression 5, table 4:3) net export ratios of engineering industries in percent

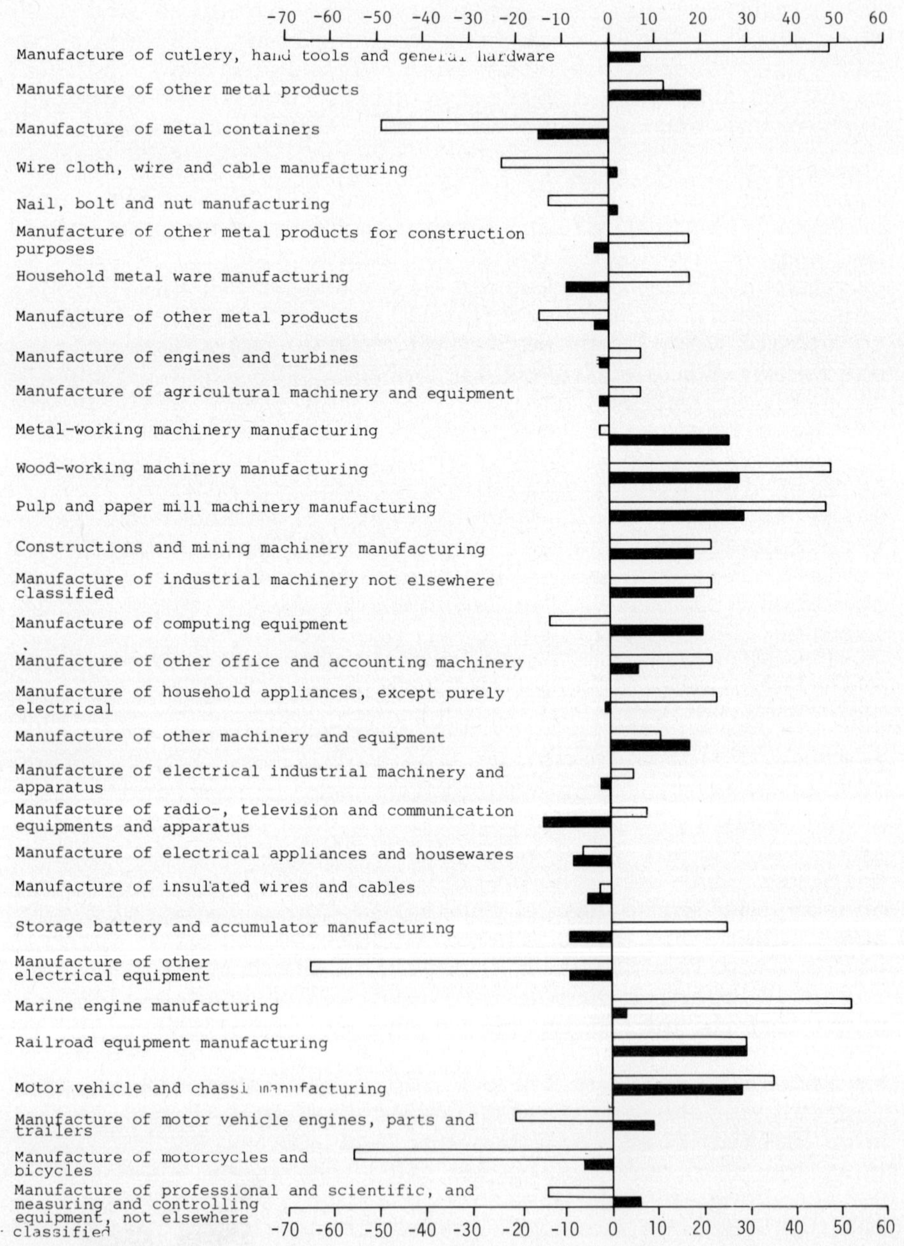

Note: The black bars show estimated net export ratios and the unfilled bars actual net export ratios.

Figure 4:2

Actual and estimated (from regression 9, table 4:3) home market shares in percent

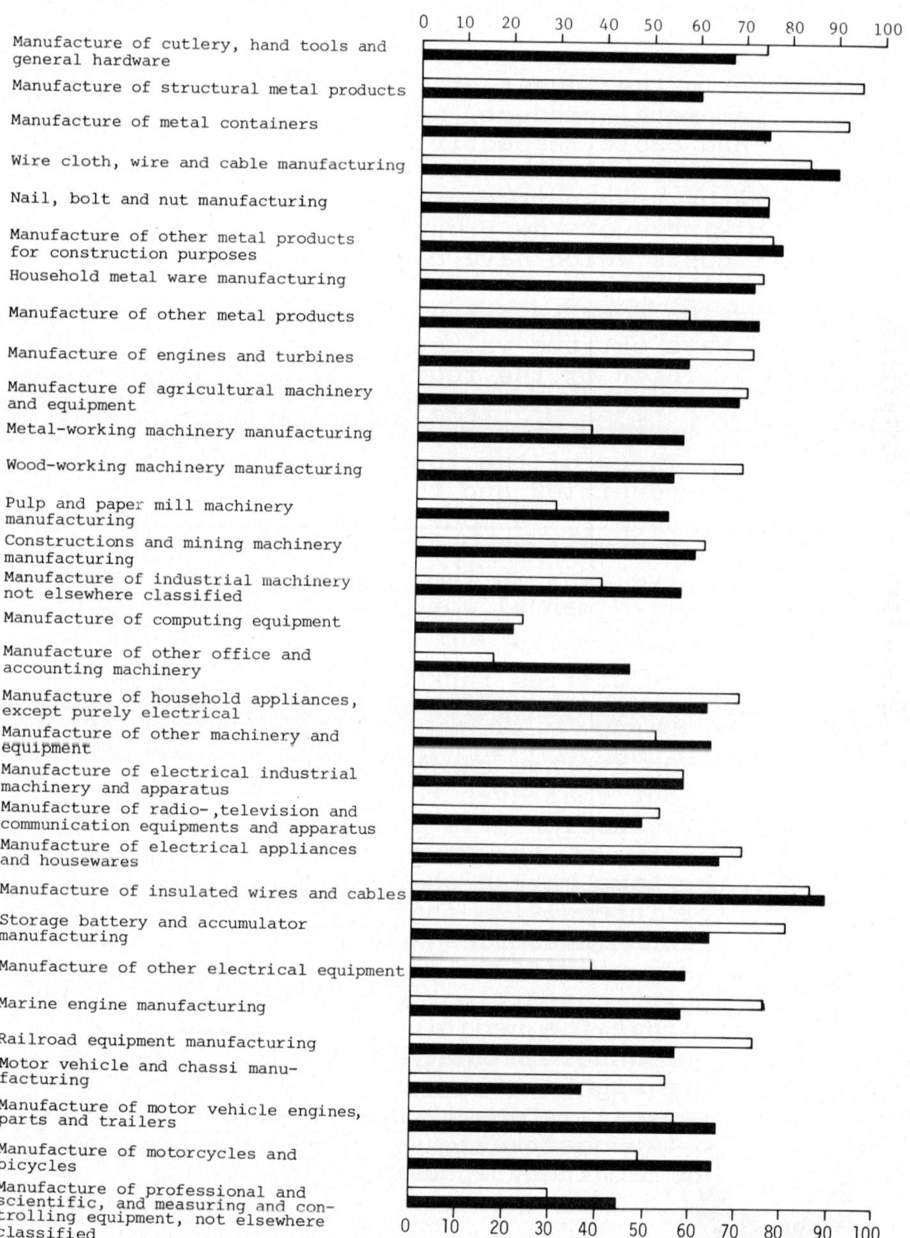

Note: The black bars show estimated net export ratios and the unfilled bars actual net export ratios.

Table 5:1

Mean value, standard deviation and relative standard deviation of commodity ton prices in 34 engineering industries in 1970

SNI no.	Industry	Mean[a] value (1000 SKr /ton) (1)	Standard deviation (1000 SKr/ton) (2)	Relative standard deviation (2)/(1) =(3)	Number of commodity groups (4)
3811	Manufacture of cutlery, hand tools and general hardware	27.4	36.3	1.32	63
3813	Manufacture of structural metal products	6.6	3.2	0.49	21
38191	Manufacture of metal containers	10.3	12.4	1.20	10
38192	Wire cloth, wire and cable manufacturing	11.4	14.1	1.24	24
38193	Nail, bolt and nut manufacturing	12.4	11.4	0.92	20
38194	Manufacture of other metal products for construction purposes	8.1	6.9	0.85	30
38195	Household metal ware manufacturing	22.6	33.0	1.46	31
38199	Manufacture of other metal products	31.0	102.3	3.30	94
3821	Manufacture of engines and turbines	35.6	35.5	1.00	15
3822	Manufacture of agricultural machinery and equipment	9.4	9.7	1.04	31
38231	Metal-working machinery manufacturing	39.9	74.4	1.87	41
38232	Wood-working machinery manufacturing	16.5	15.8	0.95	11
38241	Pulp and paper mill machinery manufacturing	13.8	4.5	0.33	6
38242	Construction and mining machinery manufacturing	10.5	5.6	0.53	18
38249	Manufacture of industrial machinery not elsewhere classified	38.1	72.0	1.89	94
39251	Manufacture of computing equipment	180.5	1.8	0.01	3
38259	Manufacture of other office and accounting machinery	68.0	58.6	0.86	31
38291	Manufacture of household appliances, except purely electrical	13.8	9.9	0.71	13
382991	Lifting and hoisting machinery manufacturing	9.8	2.8	0.29	17
382992	Liquid pump manufacturing	20.3	2.8	0.14	5
382993	Manufacture of general purpose parts of machinery	12.5	7.1	0.57	11

Cont.

SNI no.	Industry	(1)	(2)	(3)	(4)
382999	Manufacture of other machinery and equipment	23.8	19.9	0.83	67
3831	Manufacture of electrical industrial machinery and apparatus	19.8	13.4	0.68	38
3832	Manufacture of radio, television and communication equipment and apparatus	155.2	209.9	1.35	67
3833	Manufacture of electrical appliances and housewares	28.9	41.7	1.44	23
38391	Manufacture of insulated wires and cables	9.8	5.9	0.60	15
38392	Storage battery and accumulator manufacturing	18.6	27.6	1.49	9
38399	Manufacture of other electrical equipment	63.2	174.4	2.76	58
38413	Marine engine manufacturing	18.5	5.3	0.29	9
38421	Railroad equipment manufacturing	13.4	10.1	0.76	10
38431	Motor vehicle and chassi manufacturing	8.4	3.3	0.40	18
38432	Manufacture of motor vehicles engines, parts and trailers	8.5	3.2	0.37	13
3844	Manufacture of motorcycles and bicycles	18.6	7.5	0.41	13
3851	Manufacture of professional and scientific, and measuring and controlling equipment, not elsewhere classified	103.3	97.1	0.94	48
	Engineering industries	20.7	32.1	1.55	34

[a] Unweighted mean values of commodity group ton prices for individual industries. The mean value (and standard deviation) for the 34 engineering industries are calculated from the industrie ton prices, which correspond to weighted mean values.

Sources: SOS, Industri, Del 2 1970 and *SOS*, Utrikeshandel, Del 1 1970 (edited by the Swedish Central Bureau of Statistics).

These results underscore the risk for biased regression coefficients for the factor intensities at the industry level of aggregation. The first condition above that inter-industry variations in net export ratios are primarily attributable to differences in competitiveness can therefore be questioned. Thus we have to investigate whether or not the second condition holds, namely that a) the specialization relationship found in chapter 4 is not sensitive to the varying degree of heterogeneity and b) that the Swedish capital and technical personnel intensities are fairly representative or at least not seriously affected by intra-industry specialization.

Given that the expected linear relationship between the net export ratio and the factor intensities holds at the homogeneous product level, an observation tends to be more of an outlier the more technologically heterogeneous the corresponding industry is. Hence, the heterogeneity measure can be utilized for excluding relatively heterogeneous industries in an effort to investigate the fulfillment of condition 2a) above. Such an exclusion of observations diminishes the risk for obtaining biased regression coefficients in the regression on the net export ratio[9]. On the other hand, that exclusion decreases also the degrees of freedom.

5.4.2 Inter-industry specialization for homogeneous industries

In table 5:2 three regressions on the industry net export ratios are presented, one for the 16 relatively homogeneous industries, the second for 18 heterogeneous and extremely heterogeneous industries and the third for all 34 industries. The regression results for the homogeneous industries are strikingly similar to those for all industries. There is a weak specialization in skilled manual worker intensive industries in both cases. In contrast, there is no evidence that Sweden was specialized in capital and technical personnel intensive, homogeneous industries.

The conclusion is that there is so far no evidence that a large technological heterogeneity of certain industries has significantly affected the outcome of the tests presented in chapter 4 [10]. However, as discussed above the risk of non-representative industry factor intensities is large only if both the commodity factor intensities vary and the aggregation weights for the dependent and independent variables differ substantially from each other.

Table 5:2

Regressions on net export ratios of homogeneous, heterogeneous and all engineering industries 1970

Regression no.	Categories of industries	Constant	Regression coefficients (with standard errors) for K/L	L_T/L	L_Y/L_A	R^2	F-value (degrees of freedom)
1	Homogeneous industries	- 9.546	0.930 (2.799)	-0.366 (1.230)	0.599[a] (0.364)	0.19	0.960 (3; 12)
2	Heterogeneous industries	-13.461	-0.637 (2.652)	*	0.581 (0.470)	0.09	0.768 (2; 15)
3	All 34 industries	-12.733	0.268 (1.851)	-0.167 (0.853)	0.601[b] (0.292)	0.14	1.613 (3; 30)

a = significant at the 10% level (one-tail test)
b = significant at the 2.5% level (one-tail test)
* = F-level for inclusion was lower than 0.01 and the regression coefficient therefore not calculated.

Note: In order to have as many degrees of freedom as possible in the analysis, the four 6-digit industries, for which no data on L_Y/L_A was available, was given the same value of L_Y/L_A, namely the value of the corresponding 5-digit industry. A comparison with the regression results in chapter 4 where instead only the 5-digit industry was included shows that the results are strikingly similar.

5.5 *Further analysis of measurement errors*

5.5.1 *Errors due to dissimilar aggregation weights*

According to section 5.2 the measurement error in the industry factor intensity tends to be larger the larger the differences in aggregation weights for the net exports ratio and the factor intensity, respectively. The error may be small even with large differences between the two weights, if the differences tend to net out each other. However, if C_{ij}, the domestic market size, has a stronger influence on $(X_{ij}+M_{ij})$ and O_{ij} than comparative advantage then all three variables will have a similar correlation coefficient with the factor intensity[11]. In this case the measurement error of the industry factor intensity is probably not very large[12].

If, on the other hand, comparative advantage has a relatively stronger impact than market size on the weights, the correlation coefficient between the factor intensity and C_{ij} can be expected to differ from the corresponding correlation coefficients between the same intensity and $(X_{ij}+M_{ij})$ and O_{ij}, respectively. In this case, the measurement error will be relatively large at least if most of the differences have the same sign. A corollary is that the correlation coefficient between the factor intensity and $(X_{ij}+M_{ij})$ should normally differ from that between the intensity and O_{ij}.

This provides a basis for an empirical investigation of whether there are industries with relatively large measurement errors. The aim here is merely to gain further information about which observations might better be treated as extreme observations, to be excluded from an analysis of the inter-industry specialization.

An analysis of this kind requires data on factor intensities for the commodity groups. Lacking such data we have to rely again on the ton price variable as a proxy for differences in capital and technical personnel intensities. Table 5:3 presents for each industry the correlations between this variable and O, $(X+M)$, C and $(X-M)/(X+M)$ respectively. The industries are ranked according to increasing homogeneity. The table also provides the rank for each industry's capital and technical personnel intensities (from lowest to highest value) and the number of commodity groups included.

A comparison between the first three columns of correlations shows that they are practically the same for most industries. Most correlation coefficients do not significantly differ from zero. Hence, there is in general no strong indications of large measurement errors in industry factor intensities. The second subcondition for using the net export ratio, namely that intra-industry specialization should leave the Swedish factor intensity ranking relatively unaffected compared to that of the rest of the world, appears to be in general valid.

Among the extremely heterogeneous industries there are three possible exceptions to that conclusion, i.e. industries which may have large errors of measurement. They are first of all the manufacture of professional

Table 5:3

Correlation with the commodity ton prices for 32 industries 1970

SNI no.	Industry	Correlation coefficients between commodity ton prices and				Ranking from lowest to highest		Number of commodity groups
		O	$X+M$	CC	$\frac{X-M}{X+M}$	K/L	L_T/L	
		(1)	(2)	(3)	(4)	(5)	(6)	(7)
Extremely heterogeneous industries:								
3832	Manufacture of radio, television and communication equipment and apparatus	0.06	0.06	0.02	-0.02	1	30	67
38399	Manufacture of other electrical equipment	-0.05	-0.04	-0.05	-0.10[a]	4	17	58
38199	Manufacture of other metal products	-0.07	-0.07	-0.07	0.09	14	1	94
3851	Manufacture of professional and scientific, and measuring and controlling equipment, not elsewhere classified	0.13	0.41*	0.29*	0.06	3	28	48
38231	Metal-working machinery manufacturing	-0.08	-0.16	-0.16	0.21	21	19	41
38249	Manufacture of industrial machinery not elsewhere classified	-0.08	-0.07	-0.10	-0.14	15	18	94
38259	Manufacture of other office and accounting machinery	0.06	0.27	0.39*	-0.22	2	23	31

Cont.

SNI no.	Industry	(1)	(2)	(3)	(4)	(5)	(6)	(7)
3833	Manufacture of electrical appliances and housewares	-0.24	-0.21	-0.24	-0.34*[a]	7	11	23
3811	Manufacture of cutlery, hand tools and general hardware	0.71*	0.79*	0.48*	0.14	16	5	63
3821	Manufacture of engines and turbines	-0.22	-0.17	-0.23	-0.22	25	32	15
38195	Household metal ware manufacturing	-0.06	0.01	-0.08	0.16	9	2	31
Heterogeneous industries:								
38392	Storage battery and accumulator manufacturing	-0.39	-0.33	-0.37	-0.46	6	21	9
382999	Manufacture of other machinery and equipment	-0.05	-0.01	-0.10	0.06	30	25	67
38232	Wood-working machinery manufacturing	0.95*	0.89*	0.93*	0.38	22	16	11
38192	Wire cloth, wire and cable manufacturing	-0.20	-0.13	-0.24	0.12	31	8	24
3831	Manufacture of electrical industrial machinery and apparatus	-0.27	-0.26	-0.25	-0.22	19	31	38
38191	Manufacture of metal containers	-0.31	-0.26	-0.31	-0.57*[a]	12	7	10
38193	Nail, bolt and nut manufacturing	-0.24	-0.23	-0.25	-0.46*	24	3	20

Cont.

SNI no	Industry	(1)	(2)	(3)	(4)	(5)	(6)	(7)
Homogeneous industries:								
38421	Railroad equipment manufacturing	-0.02	-0.23	-0.15	0.44	28	24	10
38291	Manufacture of household appliances, except purely electrical	-0.15	-0.21	-0.16	-0.04	13	15	13
3822	Manufacture of agricultural machinery and equipment	0.01	0.05	-0.09	0.59*	18	13	31
3844	Manufacture of motorcycles and bicycles	-0.42	-0.47*	-0.46	0.03	8	10	13
382993	Manufacture of general purpose parts of machinery	0.11	0.09	0.08	-0.17	29	27	11
38194	Manufacture of other metal products for construction purposes	0.32*	0.11	0.33*	-0.01	27	4	30
38391	Manufacture of insulated wires and cables	-0.16	0.04	-0.12	-0.48*[a]	32	20	15
38242	Construction and mining machinery manufacturing	-0.26	-0.32	-0.33	0.29	23	22	18
38413	Marine engine manufacturing	0.45	0.62*	0.36	-0.38[b]	5	14	9
38241	Pulp and paper mill machinery manufacturing	-0.42	-0.07	-0.63	-0.13	26	29	6
38431	Motor vehicle and chassi manufacturing	0.01	0.25	0.03	-0.07	11	26	18

Cont.

4. Otherwise, the norm of comparison could have been a dispersion equal to zero.

5. Calculated as $(X+M)/(q_x+q_m)$ where X = export value, fob, M = import value, cif, q_x = exports in metric tons and q_m = imports in metric tons.

6. The standard deviation is a non-linear, increasing and heteroscedastic function of the relative standard deviation.

7. In the particular classification used below they will be labelled extremely heterogeneous according to the standard deviation measure but would have been labelled heterogeneous if instead the relative standard deviation had been chosen as our heterogeneity measure.

8. If instead the relative standard deviation is used, four industries will be classified as extremely heterogeneous and the other seven as heterogeneous. The classification of industries as relatively homogeneous and heterogeneous becomes the same for the two measures.

9. It has to be remembered, however, that our heterogeneity measure is intended to measure only technological heterogeneity. The products may also differ with respect to demand, transport costs or other characteristics. As might be expected there is no correlation between the Grubel & Lloyd [1975] intra-industry trade measure and the standard error of the ton prices (or for that matter the relative standard error).

10. Plots of the relationship between the net exports ratio and each factor intensity indicate that the results obtained do not depend upon where the dividing line between homogeneous and heterogeneous industries was drawn.

11. Here and in the following we assume that all possible relationships are if anything linear. Hence, correlation coefficients can be used as criteria in the analysis of measurement errors.

12. By definition $O = (C+X-M)$ if there is no stock. The size of the market will also affect X and M and the comparative advantage the sign and size of $(X-M)$.

13. Although there are some small and early producers of fine mechanic measuring devices. One early Swedish innovation was, for instance, important for the early mechanization of the car industry and for the breakthrough of the assembly lines of the Ford Motor Company.

14. The latter result is surprising since the innovation and production of hard metals and alloys require much technical personnel. One explanation may be that since firms producing these tools or machine tool parts are also large steel or alloy producers some technical personnel required for the latter might have been reported as employed in steel industry plants. If so, that would have given a negative bias for the regression coefficient of the technical personnel intensity in chapter 4.

Chapter 6

TRENDS IN SWEDEN'S INTER-INDUSTRY SPECIALIZATION 1960-1970

6.1 Outline of the analytical problem

This chapter seeks to explain the development of Sweden's engineering specialization at the industry level of aggregation. The analytical framework of chapter 4 is retained, i.e. a multiproduct-multifactor-two country model with inter-industry flows and one small economy. In contrast to chapter 4, the present chapter allows in one way or the other for changes in a) Swedish import tariffs, b) the factor intensities of the industries, c) Sweden's factor abundance, d) Swedish domestic consumption and, to some extent, e) the structure of relative (product) prices, all of which are possible determinants of the changing trade pattern. Throughout the chapter, the simplifying small country assumption is interpreted to mean that alterations in relative product prices facing Swedish producers as well as in technology are exogenously determined.

All changes are treated as if they were comparative static or long run in nature. As is evident from the above, some changes can, at least in principle, be included as independent variables in regressions on the specialization development. Even though it is impossible to say whether the accompanying trade adjustment falls completely within the period under study this provides at least some control of the time variable. That control is lost if a determinant, which cannot be explicitly introduced as an independent variable, alters over time. A change in Sweden's factor abundance is a case in point. Such a change may have fallen only partly within the period and is thus impossible to date. Analytically, we have to regard each of the years 1960 and 1970 as an equilibrium situation and assume that the trade adjustment between the two years is of the long run nature discussed in traditional factor proportions theories[1].

The trade consequences of the five types of changes mentioned above could not for data reasons be studied simultaneously in one step through a multiple regression analysis. The results of the various bits and pieces of the analysis must instead be added successively with full complexity reached only in the last sections.

6.2 The changing specialization pattern 1960-1970

The development of specialization in the Swedish patterns as measured by the net export ratio, the home market share and the export ratio (of production) between 1960 and 1970 is presented in table 6:1. The industries are ranked according to declining net export ratio changes. The net export ratio increased for 22 industries and decreased for 16. In contrast, there was a general tendency for the home market share to diminish (in 31 industries) and for the export ratio to increase (in 30 industries). That tendency was in all probability a consequence of trade liberalization among European countries.

The Swedish engineering industry has for decades held a strong international position in fabricated metal products and non-electrical machinery. In contrast it was with few exceptions not specialized in electrical machinery and appliances and transport equipment, However, several industries within the latter two sectors were among those with the largest increases in net export ratios. On the other hand many fabricated metal product and some non-electrical machinery industries experienced substantial decreases in their net export ratios.

A comparison between the two first columns of table 6:1 gives a clear impression of a negative relationship between the development of the net export ratio and its 1960 value. Since a systematic relationship of that kind may have important implications for the nature of the trade adjustment process, the next step is to investigate whether it is in fact systematic.

6.3 Is there a systematic negative specialization trend?

Figure 6:1 presents the relationship between the change in net export ratios 1960-1970 and the corresponding value of the net export ratio in 1960. The estimated regression line proves that there is a significant negative relationship. Such a relationship might have been caused by the fact that the net export ratio is a bounded variable. However, only four industries had a net export ratio lower than -40% in 1960 with the

Table 6:1

Measures of the changing industry specialization 1960-1970

The industries are ranked according to falling net export ratio changes

SNI no.	Industry	Net export ratio		Home market share		Export ratio	
		changes 1960-1970	level 1960	changes 1960-1970	level 1960	changes 1960-1970	level 1960
		(1)	(2)	(3)	(4)	(5)	(6)
3849	Manufacture of transport equipment not elsewhere classified[a]	72	-25	7	74	22	18
38431	Motor vehicle and chassi manufacturing	48	-11	3	52	21	43
38251	Manufacture of computing equipment	43	-57	-14	38	40	31
38451	Aircraft and aircraft engine manufacturing[a]	40	-82	1	57	15	7
3832	Manufacture of radio, television and communication equipment and apparatus[b]	36	-28	-9	61	25	27
3851	Manufacture of professional and scientific, and measuring and controlling equipment, not elsewhere classified	30	-46	-7	37	24	39
382991	Lifting and hoisting machinery manufacturing	28	-5	-14	68	28	30
38432	Manufacture of motor vehicle engines, parts and trailers	27	-49	-10	67	18	14
38199	Manufacture of other metal products	23	-38	-13	69	19	17

Cont.

SNI no.	Industry	(1)	(2)	(3)	(4)	(5)	(6)
38242	Construction and mining machinery manufacturing	22	1	-5	67	16	34
382992	Liquid pump manufacturing	20	-10	-24	58	33	38
38291	Manufacture of household appliances, except purely electrical	17	7	-14	84	23	18
38413	Marine engine[b] manufacturing	13	41	14	62	-9	59
38399	Manufacture of other electrical equipment	12	-77	-8	48	12	12
38249	Manufacture of industrial machinery not elsewhere classified	10	12	-6	46	10	60
38232	Wood-working machinery manufacturing	9	38	2	68	3	52
38231	Metal-working machinery manufacturing	8	-9	-14	51	18	45
3813	Manufacture of structural metal products	8	4	-1	94	2	6
38421	Railroad equipment manufacturing	6	24	-16	90	24	15
38391	Manufacture of insulated wires and cables[b]	6	-9	-6	91	6	8
38241	Pulp and paper mill machinery manufacturing[b]	5	41	-20	49	15	71
38412	Boat building and repairing[a]	3	-22	27	40	-24	49
38393	Light bulb and fluorescent lamp manufacturing[a]	0	-83	-22	62	7	5

Cont.

SNI no	Industry	(1)	(2)	(3)	(4)	(5)	(6)
382999	Manufacture of other machinery and equipment	-5	10	-5	62	3	43
38194	Manufacture of other metal products for construction purposes	-7	24	2	73	-6	38
3831	Manufacture of electrical industrial machinery and apparatus	-7	14	-16	74	13	32
38411	Ship building and repairing[a]	-11	62	-19	63	9	71
3811	Manufacture of cutlery,hand tools and general hardware	-12	59	1	72	-9	61
3822	Manufacture of agricultural machinery and equipment	-14	21	-6	76	0	33
38193	Nail, bolt and nut manufacturing	-16	2	-10	85	5	16
38392	Storage battery and accumulator manufacturing	-17	42	-4	84	-2	32
3844	Manufacture of motorcycles and bicycles	-18	-38	-30	80	12	10
38195	Household metal ware manufacturing	-20	37	-10	83	3	31
3821	Manufacture of engines and turbines	-21	29	-7	78	-1	34
38259	Manufacture of other office and accounting machinery	-25	48	-20	38	6	82
3833	Manufacture of electric appliances and housewares	-25	18	-7	78	-3	29
382993	Manufacture of general purpose parts of machinery	-30	46	-22	66	5	59

Cont.

SNI no.	Industry	(1)	(2)	(3)	(4)	(5)	(6)
38191	Manufacture of metal containers	-40	-10	-4	94	-1	5
38192	Wire cloth, wire and cable manufacturing	-42	18	-6	89	-4	15

[a] This industry is excluded from the regression analysis for reasons discussed in notes to table 4:1 and chapter 1.

[b] As mentioned in the comments to table 4:1, the 1960 levels and 1960-1970 changes are substituted by the averages 1960/1961 and 1960/1961-1970/1971 for this industry.

Note: The changes are measured in percentage units and the 1960 levels in per cent.

Sources: SOS, Utrikeshandel del 2 1960, 1961, 1970, 1971; *SOS*, Industri 1960, 1961; and Industri del 2 1970, 1971.

lowest value equal to -77%. No more than two out of thirteen industries with a negative ratio in 1960 experienced a decline in the ratio.

The highest net export ratio in 1960 was 52%. Thirteen out of twenty one industries with positive ratios showed decreases in those ratios for 1970.

The observations of figure 6:1 do not indicate that it is the boundedness of the net export ratio that has caused the negative relationship. On the contrary, the exclusion of the two extreme industries with ratios equalling -77% and 62% respectively would still give a linear and even more negatively sloped relationship.

The intercept of the regression line is positive which corresponds to an increase in the mean value of the net export ratios of the 34 industries from 4.4% to 6.5%. The net export ratio of the sum of those industries increased from -2.5% to 9.0%. The standard deviation declined from 33.3 to 28.4.

The fact that the slope coefficient of the regression line is higher than -1.0[2] implies that the negative relationship does not indicate a systematic reversal of Sweden's engineering trade specialization pattern. This leaves us with three possible explanations for the negative specialization trend:

Figure 6:1

The relationship between the 1960 levels and 1960-1970 changes of the net export ratios of engineering industries

Regression line (with standard error in parenthesis):

$y = 3.7 - 0.380x$ $R^2 = 0.297$

(0.103) $F(1;32) = 13.513$

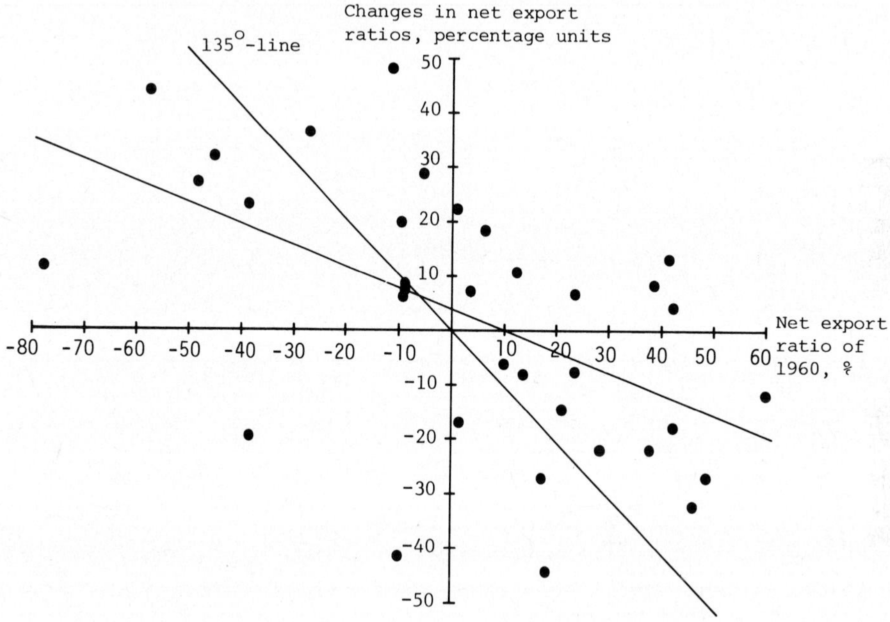

Sources: See table 6:1.

a) No relationship is found. Hence, none of the four sources of factor intensity reversals can be excluded.

b) A negative relationship is obtained implying a *systematic* tendency of factor intensities to reverse. This unlikely case would be attributable either to non-neoclassical production functions and major changes in the factor price ratio or to changing production functions.

c) A positive relationship.

The last case, which was *a priori* judged to be the most likely one, can be divided into several subcases. First, the magnitude of the explanatory value is a matter of utmost importance. A sufficiently high explanatory power means that no *major* reversals of factor intensities have occurred. Hence, an assumption of a long run stability in the factor intensity *ranking* would receive support.

Such an assumption is not restrictive enough to allow an analysis of changes in Sweden's factor abundance simply by a regression analysis between the specialization changes and the factor intensities. As was made evident in chapter 2 a stable intensity ranking is consistent with technological change which either generally increases (extensive-factor-saving bias) or decreases (intensive-factor-saving bias) the relative differences in the factor intensity. The latter case might constitute one possible explanation of the negative specialization relationship of figure 6:1.

Our more restrictive hypothesis will in consequence be: The relative differences of the factor intensity of the industries are constant in the two periods. Hence, a regression line between the indices of the factor intensity in the first and last year should not be statistically distinguishable from a 45°-line through the origin.

6.6.3 The regression results

Table 6:3 presents the regression results for the two periods and the three factor intensity measures. The results suggest a strong stability in the ranking of each factor intensity with the possible exception of the technical personnel intensity during the period 1954-1968 (regression no. 1). However, the relatively low explanatory value of that regression (63%) is attributable to one extreme observation, the computer

Table 6:3

Regressions on the 1968 factor intensity index

Index = 100 for the engineering industry as a whole.

Regression no.	Dependent variable of year		Constant	Regression coefficient (standard error)	R^2	F(1;31)
	Technical personnel intensity					
1		1954	25.0	0.825[a] (0.113)	0.634	53.688[a]
2		1959	16.7	0.824[a] (0.024)	0.975	1206.740[a]
	Capital intensity					
3	Horsepower /employee	1954	11.9	0.904[a] (0.085)	0.786	113.899[a]
4	Horsepower /employee	1959	6.2	0.978[a] (0.076)	0.843	166.337[a]
5	Electricity consumption /employee	1954	15.5	0.864[a] (0.067)	0.842	165.246[a]
6	Electricity consumption /employee	1959	16.1	0.856[a] (0.061)	0.865	197.725[a]

[a] Significant at a lower level than 1.0%.

Note: The factor intensities of a given year have been divided by the factor intensity of the engineering industry as a whole. Hence, the above regressions regress indices of 1954 (or 1959) factor intensities on the indices of 1968 factor intensities.

Source: Du Rietz [forthcoming].

If in contrast the factor price ratio changed, a similar stability might have been preserved if the elasticities of substitution were generally small and did not vary much as between the industries. In that case the resulting increase in the capital intensity of an industry would be small compared with the already large inter-industry differences in the same intensity.

Part of the observed growth in capital intensities might not be a consequence of movements along the isoquants but rather of inward shifts of the isoquants associated with technical improvement in capital equipment. However, since such an improvement creates a capital-using technical bias it did according to our results not lead to a technological bias in the sense of a systematic reduction or widening of the capital intensity differences. Either the capital-using technical bias was generally small or it tended to be of an equal relative strength in all industries.

Thus, we have obtained information about the nature of technical change in general and whether that change has introduced a systematic technological bias in the capital and technical personnel intensities. The conclusion is that there is either no such *technological* bias or a bias so small as to be negligible. The factor intensities at the end of the period can be regarded as representative for the whole period. Consequently, it should be possible to observe the influence of changes in Sweden's factor abundance in the regression coefficients for at least the capital and technical personnel intensities. As regards the skilled manual worker intensity we can only assume that the same also holds.

6.7 Factor accumulation

So far we have concluded that the observed negative relationship between the changes in net export ratios 1960-1970 and the net export ratio of 1960 can probably not be attributed to either the tariff changes alone or technological change alone. Maintaining the assumption that changes in relative world market prices are uncorrelated with the relative factor use of the industries leaves only one possible factor proportions account of the observed specialization trend: a substantial change in Sweden's factor abundance.

In a two-factor model such an (exogenously determined) change would imply different signs in the regression coefficient of the single factor intensity for the two equilibrium years studied (or for one equilibrium and the change between the two equilibria if the abundance has not been reversed but only diminished). Our case differs from that one in two important respects. First, since a multifactor model is assumed Sweden's abundance may have changed with respect to one factor compared to all others or to certain factors compared to certain others. Secondly, since it is impossible to date the possible change in Sweden's factor abundance, the two years 1960 and 1970 may not capture completely the typical differences of two equilibria.

Neither complication can be explicitly taken care of due to the lack of appropriate data. Instead, the specialization analysis makes two assumptions, one of which will be maintained during the testing of the other. Our maintained hypothesis is that Sweden's abundance of capital, technical personnel, skilled manual workers and other labour has changed so as to decrease or reverse the pre-1960 comparative advantage. Such a change can have occurred in several ways. For the first tests we shall make use of a very simple set of hypotheses. The endowments of capital, technical personnel and skilled manual workers have hypothetically increased in the 1960's. Since the net export ratio of 1970 was positively correlated only with the intensity of the last factor the hypothesis regarding its abundance in 1960 has to be slightly different. It is hypothetically assumed that Sweden was either scarcely supplied with skilled manual workers in 1960 or at least less abundantly supplied than in 1970. Moreover, Sweden is hypothetically assumed to have been scarcely supplied with both capital and technical personnel in 1960.

These hypotheses establish our multiproduct-multifactor analogue to the Rybczynski theorem (1955), namely that the changing industry specialization is non-positively related to the capital, technical personnel and skilled manual worker intensities. The inclusion of the tariff change and growth rate of consumption as independent variables is, as explained above, necessary in our more generalized model. This analogue of the Rybczynski theorem is tested next.

6.8 A regression analysis of the change in inter-industry specialization

6.8.1 Regressions on the changing net export ratio

Table 6:4 presents results of various regressions between the change in the net export ratio and the factor intensities, the tariff change and the relative change

Table 6:4

Regressions on the change in net export ratios of engineering industries 1960-1970

Regression no.	Constant	Regression coefficients (with standard errors) for: K/L	L_T/L	L_Y/L_A	$L_{Y,Q}$	Δt_{60-70}	C_{70}/C_{60}	R^2	F-value (degrees of freedom)
1	0.190	-0.031[c] (0.014)						0.125	4.559[b] (1;32)
2	-0.102		1.087[b] (0.639)					0.083	2.888[a] (1;32)
3	-0.078			0.300[a] (0.222)				0.059	1.828 (1;29)
4	-0.018	-0.024[a] (0.015)	0.821 (0.680)	0.235 (0.230)				0.216	2.479[a] (3;27)
5	0.540	-0.039[c] (0.014)	1.404[c] (0.593)	0.197 (0.193)		-5.429[c] (1.749)	-0.096[c] (0.036)	0.489	4.784[c] (5;25)
6	0.258	-0.028[b] (0.015)	1.041[a] (0.658)	0.246 (0.293)	-0.062 (1.179)	-4.188[c] (1.938)		0.343	2.609[b] (5;25)
7	0.780	-0.039[c] (0.014)	1.419[c] (0.609)	0.235 (0.264)	-0.228 (1.062)	-5.375[c] (1.800)	-0.096[c] (0.036)	0.490	3.842[c] (6;24)

[a] Significant at the 10% level (one-tail-test)
[b] " " " 5% " " " "
[c] " " " 2.5% " " " "

Legend:

K/L = a measure of the physical capital intensity, where K is the capacity of motive power (in horsepower) and L number of employees,

L_T/L = the ratio of technical personnel to total employment (both measured in physical numbers),

L_Y/L_A = the ratio of skilled manual workers to total number of manual workers (both measured in hours of work),

$L_{Y,Q}$ = a complementary measure of the skilled manual worker intensity, intended to show differences in the relative skill level of skilled manual workers, measured as the ratio between the wages per hour of skilled and unskilled workers,

Δt_{60-70} = the tariff decrease 1960-1970 on total Swedish imports,

C_{70}/C_{60} = the ratio between Swedish consumption, C, in 1970 and 1960.

in domestic consumption (or rather the ratio between consumption in 1970 and consumption in 1960). If all six independent variables are included almost 50% of the variance in $\Delta[(X-M)/(X+M)]$ is explained (see regressions 5 and 7). Almost all regressions obtain significant F-values.

The net export ratio decreased more the higher the capital intensity. Hence, the null-hypothesis of a non-positive regression coefficient cannot be rejected. Sweden appears to have become a less capital abundant country during the 1960's.

Instead its endowment of technical personnel seems to have increased according to the significantly positive regression coefficients. Thus, the null-hypothesis of a non-positive regression coefficient can be rejected as regards that intensity.

The results are not as unanimous for the skilled manual worker intensity (L_Y/L_A). At closer inspection it turned out that in regressions with both that intensity and the technical personnel intensity the latter becomes significant while the exclusion of the latter intensity brings significance to the skilled manual worker intensity. The following correlation matrix between the independent variables in table 6:4 shows a significant positive, although not particularly high correlation between the two intensities (* indicates significance at the 5% level, one-tail test):

	1	2	3	4	5	6
1. K/L	1.00					
2. L_T/L	-0.22	1.00				
3. L_Y/L_A	0.10	0.36*	1.00			
4. $L_{Y,Q}$	0.04	0.34*	0.71*	1.00		
5. Δt_{60-70}	0.16	-0.19	-0.05	-0.16	1.00	
6. C_{70}/C_{60}	-0.32*	0.23	-0.03	-0.06	0.16	1.00

In conclusion, the results may or may not lead to a rejection of the null-hypothesis of a non-positive regression coefficient. The other measures of the relative skill level of skilled manual workers ($L_{Y,Q}$) do not receive significant coefficients, i.e. the null-hypothesis cannot be rejected.

Table 6:4 shows that the net export ratio increased more the smaller the tariff decrease (cf regressions 5-7)[17]. It has not been possible to derive a theoretical interpretation of that result. As mentioned above the car industry had an initially high tariff level and ended up with a higher tariff level relative to the other industries. In order to investigate whether or not the obtained results for the tariff decrease depend on the development of that industry, it was in one regression excluded. The results were altered in two respects. First, the explanatory power of the tariff decrease was substantially reduced although its regression coefficient remained significant. Secondly, the explanatory value of the skilled manual worker intensity increased substantially, with the additional effect of making the regression coefficient for the technical personnel intensity insignificant when both intensities were included. This latter result strenghtens the case for a rejection of the null-hypothesis also for the skilled manual worker intensity.

The highly significant coefficient for the domestic market growth rate demonstrates the importance of the demand side in a factor proportions account of the changing pattern of trade specialization. The inclusion of that growth rate increases the significance of the regression coefficients for both the capital and the technical personnel intensities (cf regressions 4 and 5 or 6 and 7). Sweden tended, *ceteris paribus*, to decrease its specialization in industries with rapidly growing domestic markets. That result is contrary to what would have been expected from the original Hirsch version of the product cycle theory (see Hirsch [1967]) The factor proportions theory does not suggest any specific sign of the relationship studied but in its multiproduct-multifactor versions the rôle of the demand side is emphasized.

In conclusion, the multiproduct-multifactor-two country analogue of the Rybczynski theorem has received support However, our null-hypothesis of non-positive regression coefficients could only in part be rejected, namely for the technical personnel intensity and possibly also for the skilled manual worker intensity. The intensity of "other employees" than technical personnel and skilled manual workers should as a consequence[18] have a negative influence on the changes in Sweden's net export ratio. The by far largest groups in "other employees" are unskilled manual workers and unskilled non-manual workers (clerks, etc.). Consequently, our conclusion is that the net export ratio increased more in the 1960's the lower the intensities of unskilled workers and physical capital.

6.8.2 Regressions on the changing home market shares

Table 6:5 summarizes the regression results obtained for the change in home market shares. In contrast to regressions on both the home market share in 1970 (see chapter 4) and the changing net export ratios (see section 6.8.1) the factor intensities are completely unsuccessful in explaining the change in the home market shares. Significance is registered only for the tariff decrease. The home market share tended to decrease the larger the reduction in the tariff.

The following correlation matrix may help explain the lack of explanatory power of the factor intensities in the regressions of table 6:5:

	Home market share	
	1960	1970
Capital intensity		
horsepower/employee	0.48[a]	0.45[a]
electricity consumption/employee	0.48[a]	0.45[a]
Technical personnel intensity 1970	-0.45[a]	-0.46[a]
Skilled manual worker intensity 1972	-0.28	-0.26
Home market share in 1960	1.00[a]	0.91[a]

[a] Significantly positive at the 5% level (one-tail test).

Apparently, inter-industry differences in the home market share remain very much the same despite general reductions in home market shares. Recall from chapter 4 the evidence that varying tradability (primarily transportation costs) affected strongly the outcome of the regressions on the 1970 home market share. We might question here whether a strong stability in those tradability differences may have created a similar stability in the differences in home market shares. Some support for such a relationship is obtained from the fact that the gross trade ratios in domestic consumption were small (large) in both 1960 and 1970 for the industries which appeared to have relatively large (small) costs of transportation in chapter 4.

Again, our evidence suggests that we should rely more on the results for the net export ratio than on those for the home market share. In the remaining parts of this chapter only the latter specialization measure is therefore examined[19].

Table 6:6

Regressions on the net export ratio of 1960

Re-gres-sion no.	Con-stant	Regression coefficients (with standard error) for K/L	L_T/L	L_Y/L_A	R^2	F-value (degrees freedom)
1	-0.168	0.038[b] (0.021)			0.096	3.380[a] (1;32)
2	0.118		0.656 (0.950)		0.015	0.477 (1;32)
3	-0.049			0.278 (0.336)	0.023	0.686 (1;29)
4	-0.115	0.274 (0.241)	-0.950 (1.073)	0.355 (0.362)	0.116	1.180 (3;27)

[a] Significant at the 10% level

[b] Significant at the 5% level

Legend: See table 6:4.

export ratio is not included (cf regression 4 of table 6:4 above). To some extent at least the negative specialization trend seems to be directly associated with the factor intensities.

However, a further reduction of the negative coefficient for the 1960 net export ratio is obtained by inclusion of the tariff decrease (Δt_{60-70}) and the market growth rate (C_{70}/C_{60}) as is shown by regression 5. This is attributable to negative correlations between that ratio and the latter two variables.

Alternatively, the inclusion of the 1960 net export ratio as an independent variable can be looked upon as an effort to investigate whether or not the earlier regression coefficients of the factor intensities were specific to the functional form used. According to table 6:7 they are not.

6.9 *Summary and conclusion*

The analysis of this chapter produced four important findings concerning the relevance of the factor proportions theory. First, a striking intertemporal stability in the capital and technical personnel intensities of the industries was shown to exist for a fifteen year

Table 6:7

Further regressions on the changing net export ratio in 1960-1970

Regression no.	Constant	K/L	L_T/L	L_Y/L_A	$\left(\frac{X-M}{X+M}\right)_{60}$	Δt_{60-70}	C_{70}/C_{60}	R^2	F-value (degrees of freedom)
		Regression coefficients (with standard errors) for							
1	0.133	-0.018^a (0.013)			-0.366^c (0.107)			0.335	7.795^c (2;31)
2	-0.059		0.859^a (0.552)		-0.361^c (0.102)			0.347	8.228^c (2;31)
3	-0.097			0.407^c (0.187)	-0.383^c (0.102)			0.374	8.372^c (2;28)
4	-0.056	-0.015 (0.014)	0.507 (0.600)	0.353^b (0.203)	-0.331^c (0.106)			0.429	4.885^c (4;26)
5	0.396	-0.029^c (0.013)	1.036^b (0.547)	0.302^b (0.177)	-0.261^c (0.097)	-4.905^c (1.574)	-0.066^b (0.034)	0.608	6.213^c (6;24)
6	0.037				-0.380^c (0.103)			0.297	13.513^c (1;32)

[a] Significant at the 10% level
[b] " " " 5% "
[c] " " " 2.5% "

Legend: $[(X-M)/(X+M)]_{60}$ = net export ratio of 1960. See also legend to table 6:4.

Note: $[(X-M)/(X+M)]_{60}$ has in all regressions been forced to enter as the last independent variable.

period. Moreover, that stability held not only for the factor intensity *rankings* but also for their *relative* differences.

This being the case, it should not come as a surprise that factor intensities proved to have good explanatory powers in regressions on changes in the net export ratio. As suggested by multiproduct-multifactor models, changes in trade impediments and the existence of non-homothetic demand patterns may play a significant rôle for the movement of trade over time and possibly distort the factor intensity relationships. The former was found to be true but the latter was not the case.

The last finding was the negative relationship between the net export ratio of 1960 and its changes up until 1970. Moreover, the sign of the coefficients for the three factor intensities against both that ratio and its change suggested that a change might have occurred in Sweden's factor abundance from a capital towards a technical personnel and skilled manual worker abundance. Although the explanatory value of the factor intensities was too small to account for more than a minor part of the observed negative specialization trend, the result is sufficiently clear to serve as a basis for our further efforts to study the changing trade patterns.

Another finding which in principle might help to explain the negative specialization trend was the observed, although relatively small reductions of the relative inter-industry differences in the technical personnel intensity. This explanation requires a comparative advantage in technical personnel intensive production and an exogenously given technological change. As mentioned we have found an increased specialization in such production. However, that very finding puts in question whether the observed declining tendency was exogenous since a loss in comparative advantage of technical personnel extensive industries would cause them to hire more technical personnel in order to facilitate changes in the product mix, the mechanization of production and perhaps also to develop new products. In that case the declining tendency would itself have been caused by the altered Swedish abundance. Depending on what measures were taken by the new technicians in the technical personnel extensive industries the adjustment process might have either been delayed or prompted.

The interpretation of a changing Swedish factor abundance rests on the small-country assumption of a development of world market prices which is uncorrelated with the factor intensities of the industries. Suppose this was the case. Then the change in Sweden's abundance of human capital must have come about through an increasing endowment since the factor price ratios remained constant. That conclusion relies in turn on an assumption of no factor market distortions, e.g. factor price rigidity. But the increased endowments of human capital would not in the latter case have been absorbed unless *either* a) an initial excess demand for human capital existed in 1960, *or* b) the domestic tariff decreases meant a favourable price development of human capital intensive goods as domestic market prices approached the existing world market prices *or* c) the world market prices increased relatively for such goods.

As shown in Ohlsson [1973] chapter 2 an excess demand for both technical personnel (or rather engineers) and skilled manual workers did prevail at ruling wage rates. The former market reached a balance in the second part of the 1960's while the latter has not yet reached such a balance. Both disequilibria suggest that factor market differentials might have existed in the 1960's of a kind putting the neoclassical assumptions of perfect competition with free entry out of order.

As regards the second condition above it is related to the old-time Heckscher [1919] (and Stolper-Samuelson [1941]) issue of which factor is protected by tariffs and their possible impact on factor returns. As shown in the chapter the size of the tariff decreases were negatively, but insignificantly correlated with the three human capital intensities. If anything that would have meant a protection of the relative wages of *unskilled* labour, which conclusion is in fact receiving some support from Lundberg [1976]. The relevance of the third condition will be discussed further below.

As regards the weakening of Sweden's earlier abundance in non-human capital, the following can be said. Recall first that no engineering industry is more capital intensive than the manufacturing industry as a whole and that the net export ratio of 1970 was uncorrelated with that intensity. Hence, Sweden might even as late as 1970 have had a certain comparative advantage in capital intensive *manufacturing* goods.

Results of other studies and the increasing capital/labour ratios of the subindustries suggest that the rent/wage ratio has decreased in Sweden. This was probably true for other industrial countries as well. In fact, since Sweden decreased its specialization in capital intensive industries the rent/wage ratio should have declined less in Sweden than abroad assuming again perfect factor markets. The increased domestic endowments of non-human capital sufficed to bring about an increased capital intensiveness of the industries at the cost of a decreased emphasis on the most capital intensive ones[22]. Such a development would have been stimulated by *either* a) an initial disequilibrium in 1960 between the existing rent/wage ratio and the capital/labour ratio in each industry *or* b) trade liberalization such that the Swedish rent/wage ratio fell in absolute terms but increased compared to other countries *or* c) a fall in the world market prices for capital intensive goods relative to other goods.

It is difficult to evaluate the first condition despite the general view that an excess demand existed in Sweden for financial capital throughout the 1950's and 1960's. Hence, various rationing systems were in use to distribute that capital. Consequently, we have reason to believe in factor market differentials affecting the structural adjustment of firms differently.

The second condition is, in part at least, contradicted by the fact that Swedish tariffs were higher, the lower the capital/labour ratio of the products (see above and also Carlsson & Ohlsson [1976] and Lundberg [1976]). Accordingly, the trade liberalization in Sweden was associated with an upward pressure on the domestic rent/wage ratio. Condition b) above could in consequence only hold if foreign tariffs protected capital rather than labour and if the subsequent trade liberalization in the 1960's brought about such a large fall in the rent/wage ratio that it led to a relative price decrease for capital intensive products.

Thus, much more conclusive results could have been gained with data on world market prices for the industries in 1960 and 1970. Even without those data some more insight may be gained. One cause of a changing relative price situation towards increases for human capital compared to non-human capital intensive production is demand growth. Recall that Sweden and the rest of the world were assumed to have identical demand functions with income elasticities which are constant over the relevant income range. Then the significant negative correlation between the relative growth of domestic consumption and the capital intensity (-0.32) together with its positive, although insignificant correlation with the technical personnel intensity (+0.23) might suggest one mechanism behind such a relative price change. The market growth rate is also significantly, positively correlated with the metric ton price of Swedish exports and imports (0.45 respectively 0.46), which variables according to chapter 5 varied positively with the technical personnel intensity and negatively with the capital intensity. Assuming that the latter relationship also holds for the metric ton price of OECD exports (P_E), the following regression for 106 commodity groups gives further indications:

$$P_{E,70} = \underset{(0.325)}{0.143} + \underset{(0.042)}{1.211}(P_{E,64}) \qquad R^2 = 0.888 \quad F(1;104) = 825.764$$

Since the slope coefficient deviates positively from 1.0 the ton prices of OECD exports have risen more for technical personnel intensive, capital extensive products than for technical personnel extensive, capital intensive ones. This development might have been generated either by a relative price rise of the former products or by a relatively larger increase in their technical personnel intensities or by a relatively larger decrease in their capital intensities. The latter two tendencies did at least not hold for Sweden.

There are in consequence indications, although admittedly weak ones, of world market price development in favour of technical personnel intensive and capital extensive production. Such a change transmitted to Swedish producers might possibly explain the observed changes in both specialization and technology, but why did the Swedish wage ratios remain unaffected? It appears that the trade adjustment with respect to skilled labour had to come about through demand generated changes in Sweden's endowments. The nature of the formation of such skills suggests that the adjustment process had to be slow.

The next chapter takes a look at the part of the adjustment process which comes about through inter-industry differences in the development of new, permanent and exiting firms. The intention is to gain further insight into the nature of that adjustment in an economy which appears to have had some factor market differentials.

Footnotes

1. The type of intermediate adjustment pattern discussed in Mayer [1974] is not studied here. However, chapter 7 focuses on firm adjustment patterns whose analysis resembles the medium run problems of Mayer.
2. The slope coefficient of the 135°-line in figure 6:1.
3. As well as direct wage costs.
4. Since Sweden's relative wages were constant and the wage gap between Sweden and other countries were more or less constant for manual workers.
5. As well as with the other independent variables of the regressions.
6. Note the relative price assumption means that changes in domestic price structures are accomplished by the tariff decreases which are represented by another independent variable in the regressions.
7. The Swedish industrial statistics before 1967 published the necessary data for only five engineering industries.
8. Significant at the 5% level.
9. Since both Volvo and SAAB could import British parts which had no tariffs after 1967, the effective tariff rate might even have increased in contrast to the nominal ones.
10. Cf for instance Herberg, Kemp & Magee [1971] or the recent monograph Magee [1976].
11. See Minhas [1962], Yeung & Tsang [1972] as well as a recent regional application by Buck & Atkins [1976].
12. Hence the risk for reversals due to intra-industry specialization is substantially lower.
13. From Du Rietz [1975] and [forthcoming].
14. Or thirty-four when the skilled manual worker intensity is not included as an independent variable.
15. At least in part this fact is attributable to the rapid electrification of the machinery equipment during recent decades. The correlation coefficient between the two measures is 0.65.
16. Or rather the members of the Swedish Engineering Employers' Association (Sveriges Verkstadsförening).
17. It may be noticed that the simple correlation coefficient between the two variables is not significant (-0.22). Thus, the estimation of the structural influence of the tariff decrease has to include variables which might reflect differences in elasticities of production and consumption.

18. Since those three categories related to total employees sum up to 100%.

19. See, however, chapters 8 and 9 and in part also chapter 7 for additional results on the home market share and related specialization measures.

20. That hypothesis is also based on our knowledge of the regression coefficient for 1970, which coefficient was not significantly different from zero.

21. That the latter two intensities were given the same null-hypothesis was motivated by the fact that the skilled manual worker intensity obtained a weaker positive relationship with the changing net export ratio than the technical personnel intensity with the opposite result for the net export ratio in 1970.

22. This does not only hold for the engineering industry, since that industry has rapidly increased its share in exports and imports and at the same time increased its absolute and relative export surplus.

Chapter 7

FACTOR MARKET BARRIERS AND OTHER BARRIERS TO ENTRY AND EXIT[1]

Co-authored by Gunnar Du Rietz

7.1 Outline of the analytical problems

Traditional factor proportions theories rely on certain strategic assumptions of the classical static equilibrium theory. That theory presupposes that all firms have identical U-shaped, long run cost curves with the same minimum points. There are no external effects and the industry has constant costs due to the absence of barriers to entry (and exit) and a large number of small, potential producers. Increased demand in the very short run is met by existing producers who are stimulated to increase production by the increased market price. In the medium run the production increase is shared by new and existing firms. Shares in production increase are determined by the supply elasticities of existing firms and the speed with which new firms can start up and expand production.

In an open economy a third source of supply to the domestic market is imports. Hence, the adjustment of domestic (existing or potential) producers is affected also by the foreign supply elasticity. In accordance with the small-country assumption of earlier chapters we shall assume perfectly elastic foreign supply (and demand) curves. Thus, an outward shift in the domestic demand curve would lead to an equivalent, short run increase of imports in the absence of entry barriers for foreign producers. Furthermore, new domestic firms would have no incentives to enter the market. For any given product the medium and long run production response of existing or potential domestic producers will depend upon whether or not the home country has a comparative cost advantage. If it does, new producers will enter and absorb the whole (or at least a large part) of the increased domestic market and cause indirect general equilibrium adjustments in the production of other goods until an equilibrium is reached with balanced trade.

The outlined "ideal" case is not a very likely one. Barriers to entry and exit may exist for new domestic and foreign producers through commodity market or factor market barriers whether autonomous or policy imposed[2]. Such barriers are gaining increasing recognition in pure trade theory, especially in its welfare branch[3] and have for long been a focus of interest in industrial economics although usually not in a general equilibrium setting.

Our purpose is to investigate the rôle of certain entry and exit barriers for domestic and foreign firms selling, at least potentially, in the Swedish market. Two independent, exogenous causes of change in internal specialization of these firm categories are recognized, namely a) the supply incentives generated by the assumed alteration in Sweden's comparative advantage and b) the demand growth which is assumed to be in accordance with non-homothetic, for the two countries identical demand curves with constant income elasticities. The growth of income is presumably created by economic growth due to the effects of trade liberalization, technical improvement and factor accumulation. Consequently, we are broadly maintaining a factor proportions framework although oriented more towards adjustment problems of a more medium run than toward the very long run nature[4].

For this reason a source other than Swedish industrial statistics has to be utilized (see Du Rietz [1975] and [forthcoming]). The use of that source forces us to switch the period from 1960-1970 to 1959-1968 and to make some other changes which may affect comparability with earlier chapters.

7.2 *Delineating specialization measures for categories of firms*

Recall from section 2.6 the identity $O_k \equiv C_k + X_k - M_k$ where industry $k = 1,\ldots,n$, O = domestic production, C = domestic consumption, X = exports and M = imports. Total production (O_k) and exports (X_k) of domestic producers can be divided into production and exports of various categories of domestic firms, defined as follows:

a) *New firm entry* is defined to include first of all a firm which in 1960-1968 establishes a new industrial plant, and which is not owned by another firm in the same or other industries. The firm must still exist in 1968. Secondly, a new firm entry may also be a plant of an independent firm which has changed its output mix so much during the same period that it could be classified as belonging to a new industry[5]. The category new firm entry is denoted *N1*.

b) *Diversification entry* is for the same period defined as a plant established by another industrial firm with production in another industry. It may be a completely new plant or a plant earlier used to produce goods belonging to another industry. Diversification entries (henceforth labelled *N2*) are not always hindered by the same entry barriers as independent entries. For a firm starting a new production unit in the same industry as its old plant(-s) the new unit is called an expansion entry and is grouped together with permanent firms.

c) *A permanent firm* is a firm with production both in 1959 and 1968 in a given industry, usually also at the same plants. This firm category (henceforth labelled *P*) also includes contraction exits.

d) *Independent exits* are either independent firms which have discontinued their total operation or re-organized their production so that the firm (plant) is reclassified into another industry. Independent exits (henceforth labelled *E1*) are thus defined analogously as independent entries. Such exits had production in 1959 but not in 1968.

e) *Specialization exits* consist of discontinued or reclassified plants owned by firms with production also in one or more other industry(-ies). Hence, they compare with diversification entries and will be labelled *E2*.

Those definitions correspond to the data material provided by Du Rietz [1975] and [forthcoming]. Rewriting the identity above for the two years 1959 and 1968 (index *59* and *68*, respectively) and separating the five groups of domestic firms we have for the industry k:

$$C_{k,59} \equiv P^{O}{}_{k,59} - P^{X}{}_{k,59} + E1^{O}{}_{k,59} - E1^{X}{}_{k,59} + E2^{O}{}_{k,59} - E2^{X}{}_{k,59} + M_{k,59} \quad (1)$$

$$C_{k,68} \equiv {}_{P}O_{k,68} - {}_{P}X_{k,68} + {}_{N1}O_{k,68} - {}_{N1}X_{k,68} +$$

$$+ {}_{N2}O_{k,68} - {}_{N2}X_{k,68} + M_{k,68} \tag{2}$$

Since it is impossible to obtain data on exports for any one of the five firm categories their exports are lumped together and labelled X_{59} and X_{68} respectively. Both identities above are also divided through by the domestic consumption, C, for the respective year. Subtracting the resulting expression (1) from the resulting expression (2) and rearranging yields:

$$\left[\frac{{}_{P}O_{k,68}}{C_{68}} - \frac{{}_{P}O_{k,59}}{C_{59}}\right] + \left[\frac{{}_{N1}O_{k,68}}{C_{68}}\right] + \left[\frac{{}_{N2}O_{k,68}}{C_{68}}\right] -$$

$$- \left[\frac{{}_{E1}O_{k,59}}{C_{59}}\right] - \left[\frac{{}_{E2}O_{k,59}}{C_{59}}\right] \equiv \left[\frac{X_{68}-M_{68}}{C_{68}} - \frac{X_{59}-M_{59}}{C_{59}}\right]. \tag{3}$$

Each one of the first five brackets denotes the change in *production* specialization of a firm category. Their changes sum up to the change in Sweden's international specialization, here measured as $\Delta[(X-M)/C]$ instead of $\Delta[(X-M)/(X+M)]$[6]. For the sake of simplicity let us denote the six brackets of formula (3) as follows by the expression (dropping the k index)

$$\Delta\left(\frac{O_{P}}{C}\right) + \Delta\left(\frac{O_{N1}}{C}\right) + \Delta\left(\frac{O_{N2}}{C}\right) + \Delta\left(\frac{O_{E1}}{C}\right) + \Delta\left(\frac{O_{E2}}{C}\right) \equiv$$

$$\equiv \Delta\left(\frac{X-M}{C}\right). \tag{3a}$$

Observe that $\Delta(O_{E1}/C) = -({}_{E1}O_{59}/C_{59})$ and that $\Delta(O_{E2}/C) = -({}_{E2}O_{59}/C_{59})$. These definitions are chosen so as to obtain hypotheses regarding the sign of the regression coefficients which are similar for the firm categories in the absence of entry and exit barriers etc.

Recalling the hypothetical case of no distortionary barriers to entry or exits (apart from tariffs) discussed in section 7.1, we have the following hypotheses based on the results of chapter 6 for a medium *or* long run period consistent with our earlier findings for $\Delta[(X-M)/(X+M)]$:

1. $\Delta(O_p/C)$, $\Delta(O_{E1}/C)$ and $\Delta(O_{E2}/C)$ are all non-positively related to the capital intensity, non-negatively to the two human skill intensities and non-positively to the tariff decrease and market growth rate.
2. $\Delta(O_{N1}/C)$ and $\Delta(O_{N2}/C)$ are both negatively related to the capital intensity, positively to the two human skill intensities and negatively to the tariff decrease and market growth rate.
3. Since the production specialization changes of domestic firms will sum up to a value equivalent to the changing trade specialization (see formula 3a) the latter variable $\Delta[(X-M)/C]$ is assumed to be negatively related to the capital intensity, positively to the two human skill intensities and negatively to the tariff decrease and market growth rate.

Those three classes of hypotheses can be utilized as norms of comparison for the analysis of entry and exit barriers. The following section presents the various types of barriers which are empirically investigated later in this chapter.

7.3 *Barriers to entry and exit*

In principle, we may distinguish barriers primarily affecting factor markets from those affecting commodity markets, barriers which are imposed by policy from those which are not (cf Bhagwati [1971]), and lastly, barriers which exist at home from those abroad. However, the small country assumption implying (perfectly) elastic foreign supply and demand elasticities allows us to assume away *foreign* barriers. That leaves us with four types of domestic barriers to entry and exit, respectively, to be discussed in the following two sub-sections.

7.3.1 *Factor market barriers*

Two types of factor market barriers distinguished above were endogenous and policy-imposed ones. We shall not analyse the latter barriers because of the indirect and non-measurable nature of these incentives[7]. There are two kinds of endogenous factor market barriers of possible importance. First, there are factor market differentials directly connected with the nature, timing and spread of Sweden's changing factor abundance.

Secondly, there are barriers which are not at all associated with that change but instead were created by "regular" technology characteristics, factor price rigidities etc. This is not the place for a deepgoing study of the latter type of barriers[8], since we are primarily interested in barriers which can systematically divert the domestic adjustment processes for permanent, new and exiting producers from the patterns hypothetically described above.

The reason for studying the first type of barriers is that it is impossible to date the change in Sweden's factor abundance. Hence, the change may have come about gradually during the 1960's in such a way as to create resource constraints or factor price differentials for certain groups of producers. The long run rigidity in Swedish wage ratios[9] discussed in chapter 6 may also mean that there have been no major differences between the relative wages of such producer groups. If so, the differences could reflect the conditions of accessibility to scarce resources. The conclusions of chapter 6 were that non-human capital had become more scarce and human capital more abundant during the 1960's. In the following, we shall discuss hypothetically why the five groups of domestic producers defined in section 7.2 might have had different access to non-human and (the two kinds of) human capital. Permanent producers are used as our norm of comparison.

Resource constraints with regard to *non-human capital* are most likely financial constraints limiting the availability of financial capital, which probably affects new firm entries (the *N1* category) more than both permanent producers (*P*) and diversification entries (*N2*). The excess demand for capital on the domestic market at going interest rates, credit rationing by the Riksbank etc give reason to such a hypothesis. If anything, we should expect new firm entries to be fewer and/or grow slower in capital intensive industries in comparison with the growth of permanent producers and of diversification entries. Very large differentials in access to financial capital would imply that producers existing already in 1959 might even have increased their specialization in capital intensive industries. If in contrast those producers also felt a shortage of financial capital, the following relationships might be expected to hold: diversification entries $[\Delta(O_{N2}/C)]$, the comparative development of permanent firms $[\Delta(O_{P}/C)]$ and both independent exits $[\Delta(O_{E1}/C)]$ and specialization exits $[\Delta(O_{E2}/C)]$ are negatively related to K/L. These relationships are more likely to hold, if in addition the relative world market price of capital intensive products fell, since producers already in the market in 1959 would probably notice such a tendency earlier than potential producers.

Resource constraints with regard to *technical personnel* and *skilled manual workers* are more difficult to analyse because very little is known about the functioning of the markets for those factors in the medium to long run. Two important facts should be emphasized. One is the fact that the excess demand for engineers (≈ technical personnel) gradually disappeared during the period of study while the excess demand for skilled manual workers did not. If anything, that should imply diminishing entry barriers to producers intensive in technical personnel but not to producers intensive in skilled manual workers.

Secondly, a possibly important determinant on the supply side is the different career opportunities for technical personnel and skilled manual workers. The career opportunities of the latter group are relatively small as long as they remain manual workers. Especially during the 1950's, the existing shortage of engineers created opportunities for skilled manual workers to increase income by becoming technical personnel after, for instance, some engineering courses of instruction by correspondence. The same situation meant excellent career opportunities for educated engineers *within* existing firms, where opportunities should have gradually diminished as the excess demand for engineers in the 1960's disappeared.

Apart from this, it is difficult to evaluate how an increased abundance of technical personnel and skilled manual workers might have affected certain groups of producers. Since the initial specialization pattern was negatively correlated neither with L_T/L nor L_Y/L_A there was apparently a substantial production in some industries with high values for the two intensities. Hence, the increased specialization on technical personnel and skilled manual worker intensive production need not have been associated with the establishment of new plants (categories *N1* and *N2*).

By analogy with the arguments behind possible differencies in the access to financial capital, the relationships between the two human skill intensities and the five specialization measures may be taken to indicate differences in the strength and functioning of possible resource constraints in the market for these production factors. In conclusion, it is not possible to formulate a specified hypothesis for each one of those relationships as regards the sign of the regression coefficients for the technical personnel and skilled manual worker intensities.

The *second kind of endogenous factor market barrier* was that which could not be associated with the changing factor abundance of Sweden. As indicated by the surveys of Magee [1973, 1976] there are many possible factor market differentials which can work as entry or exit barriers for domestic producers. Since our purpose is limited to the study only of those barriers which might have systematically affected industries with certain technologies we may restrict ourselves to those which are obviously associable with technology differences.

Bain [1956] distinguished three entry barriers, namely a) the extent of production economies of scale relative to the size of market, b) product differentiation, and c) absolute cost advantages for existing firms. Recall that our study differs from the framework of Bain [1956] in two important respects. First of all, Sweden was a small open economy already in 1959 and secondly we are primarily studying distortionary or delaying factors to an adjustment of the composition of trade and production under the assumption of a changing factor abundance.

Whether and how the size of the domestic market combined with economies of scale determines the pattern of trade specialization of a country has been discussed by Burenstam Linder [1961] and Drèze[10]. It was, however, not possible to construct a methodology for application at the industry level on the basis of their tentative arguments. Some efforts are made in chapters 8 and 9 to test whether the relative size of the domestic and world market had any structural impact on the pattern of Swedish commodity group specialization. Suffice it to say that the analysis gives us no reason to believe that an omission of an "economies of scale relative to the size of the market"-variable would seriously affect the conclusions of this chapter.

The second of Bain's entry barriers was product differentiation. Product differentiation has earlier been discussed by Grubel & Lloyd [1975], Hufbauer [1970], Ohlsson [1973] and Wells [1972] with respect to possible trade impact. It is examined in chapters 8 and 9 of the present study. Neither the earlier studies nor the present examination indicate that the pattern of product differentiation can be easily explained and associated with the trade pattern in a systematic way. Despite this the two heterogeneity measures of table 5:1 were tried together with other independent variables in regressions on firm adjustment patterns. It was assumed that the more technologically heterogeneous an industry is the easier it would be to enter that industry by specialization on certain more or less odd products. However, no significance was obtained for either heterogeneity measure[11]. Therefore, those variables were omitted from the final regressions and will henceforth be ignored.

The third class of Bain's entry barriers is absolute cost advantages for existing firms. Such advantages may be attributable to a large amount of capital required to operate a plant of minimum efficient scale or to operate big R & D or sales divisions leading to technological or market gaps etc. However, we had no indications that neither the R & D nor sales personnel intensities had any systematic impact on the Swedish trade pattern. Since they were also impossible or alternatively difficult to analyse at the industry level, the two kinds of absolute cost advantages are not studied in what follows.

Absolute cost advantages attributable to varying capital requirements of plants of minimum efficient scale were according to Du Rietz [forthcoming] an important entry barrier. New firm entries were *smaller* in industries with relatively larger initial (average) capital stocks per plant (K/P of 1959). That study also includes metal and plastic product industries which implies a much larger variation in K/P than that obtained within the engineering industry. The hypothesis is here that new firm entries $[\Delta(O_{N1}/C)]$ should be negatively correlated with the capital stock per plant among the industries. For the same reasons that variable should also reflect barriers to exit for producers with production in 1959. Hence, the comparative development of permanent $[\Delta(O_P/C)]$ and exiting $[\Delta(O_{E1}/C)$ and $\Delta(O_{E2}/C)]$ firms is hypothetically more favourable in industries with a large capital stock per plant.

7.3.2 *Commodity market barriers*

By analogy with factor market barriers, commodity market barriers can be classified as either policy-imposed or endogenous. The results of chapter 6 emphasized tariff decreases as important in determining in which markets foreign producers could increase their exports to Sweden. Hence, a tariff decrease can be looked upon as a removal of a policy-imposed exit barrier for existing domestic producers. Depending upon how the initial barriers affected the domestic price level, profits, and volume of domestic production, the tariff decrease may or may not have had an impact on new firm entries after 1959. The tariff decrease will be tested as an independent variable also for entries although without any specific hypothesis regarding the sign of the regression coefficient.

As mentioned in earlier chapters it was impossible to measure both non-tariff trade barriers and natural trade barriers such as transport costs. The fact that transport costs impedes exports and imports relative to the size of domestic production and consumption emphasizes the need to pay special attention to whether or not using domestic consumption in the denominator of the specialization measures biases the results. According to chapter 4 this bias may be expected to appear for the regression coefficients of the capital and skilled manual worker intensities[12].

It is possible to take account of two endogenous barriers to entry and exit. One is based on the assumption that cartels or collusion between existing, domestic producers[13] may prevent or delay new domestic and foreign producers from entering the market. Permanent firms are hypothetically assumed to have a comparatively rapid production increase in such markets while exits are assumed to be fewer (or of smaller size). Cartels and collusions are usually expected to be more common and/or effective the fewer the number of domestic firms and the larger their market share. In a small, relatively open economy they are presumably more effective the smaller imports are. Therefore we have chosen to include as an independent variable a four-firm concentration ratio of the year 1959 which is adjusted by multiplication with $1 - (M/H) = H/C$. That ratio will be labelled *CONC*.

One reason for a large value of *CONC* in 1959 may be large tariffs. The trade liberalization of the 1960's should have led to diminished seller concentration in tariff-protected markets, whose effect was probably reflected in the regression analysis by the above-mentioned tariff decrease variable. An inter-industry stability in relative transport costs suggests that high concentration ratios due to such costs could have remained effective during the period. There may be a number of other reasons for highly concentrated sellers markets, for instance large economies of scale relative to the size of the domestic market, large minimum efficient scale of production, etc but those determinants do not generally prevent *foreign* producers from exporting to Sweden. Therefore they do not establish sufficient conditions to create entry barriers for imports although they may do so for new domestic producers.

The second endogenous commodity market "barrier" is the growth rate of the market, here measured by the growth rate of the domestic consumption ($\Delta C/C_{59}$). It creates incentives or disincentives to enter and exit certain industries compared with others. The results of chapter 6 showed that Sweden's specialization diminished in industries with fast-growing domestic markets. Hence, foreign producers have increased their specialization in these industries.

As discussed by Du Rietz & Hause [1976] and Du Rietz [forthcoming] a rapid growth of demand stimulates the growth of production of new firms. In their dynamic model the marginal cost of capacity of permanent firms is assumed to be an increasing, convex function of their growth rate of production. New firm entries are assumed to have marginal cost of capacity functions which are linear or concave. The more rapid the market grows the easier it is for new firms to enter it. The comparative growth of new firms [$\Delta(O_{N1}/C)$ and $\Delta(O_{N2}/C)$] will then be faster for two reasons namely 1) more new firms enter the market and 2) their production tends to increase more. However, our analysis here is based on a small open economy, comparative static model. The equilibrium is attained by either new firm entries or imports. The traditional adjustment mechanisms of such static models do not give reason to assume a given functional form of the relationships with the market growth rate. Therefore we shall have to let this question be answered by the empirical analysis.

Among the other reasons for expecting lower entry barriers in rapidly increasing markets are rapidly decreasing absolute cost disadvantages, less favourable conditions for upholding cartels or collusion etc. However, as discussed above the influence of those determinants are hopefully absorbed by other independent variables.

According to the above, foreign producers have in fact increased their specialization in fast growing markets and new domestic ones follow hypothetically the same pattern. Moreover, exits [$\Delta(O_{E1}/C)$ and $\Delta(O_{E2}/C)$] can be assumed to be smaller the larger the market growth rate. Hence, we would expect permanent producers [$\Delta(O_P/C)$] to decrease their specialization in products with rapid market growth.

7.3.3 The norms of comparison

Expression (3a) above included six measures of changing specialization of which five measured the change attributable to new firm entries, diversification entries, permanent firms, independent exits and specialization exits. If each one of the six measures is run against each one of the above specified independent variables in linear regressions the sum of the regression coefficients over the six regressions should be 1.0. For two reasons this does not in fact hold.

One reason is that the estimates of X, M and C are derived from classifications of products (SITC-Revised and the Brussels nomenclature) while O_P, O_{N1}, O_{N2}, O_{E1} and O_{E2} are all obtained from an industry classification of each plant. Since many plants produce products belonging to several industries the sum of O_P, O_{N1}, O_{N2}, O_{E1} and O_{E2} will not be equal to $(C+X-M)$. Hence, $\Delta[(X-M)/C] \neq \Delta(O/C)$ if O = total production calculated from plant data. Thus, we have two problems in comparing the regression results with the results of chapter 6, the first being the use of C in the denominator instead of $X+M$ and the second that $\Delta[(X-M)/C] \neq \Delta(O/C)$. As a consequence we shall not use $\Delta[(X-M)/C]$ as our norm of comparison in commenting upon the existence of entry and exit barriers but instead $\Delta(O_P/C)$, i.e. the comparative development of permanent firms.

The abandonment of $\Delta[(X-M)/(X+M)]$ as our norm of comparison does not create severe difficulties. The purpose of the analysis is merely to find out whether entry and exit barriers have distorted the adjustment of trade and production following a change in Sweden's factor abundance compared with a situation of perfectly functioning commodity and factor markets. That purpose is fulfilled by investigating whether or not the various groups of firms differ from each other first of all with respect to the regression coefficients for the factor intensities. Secondly, even if that is not the case other entry and exit barriers may have led to distortions from the "ideal" pattern which are not related to the factor intensities but which can help to explain the fact that those intensities did not account for even half the variance of $\Delta[(X-M)/(X+M)]$ in chapter 6. The effects of factor market barriers can be distinguished from the effects of commodity market barriers only if the factor intensities are not highly correlated with the other independent variables. Table 7:1 shows the relevant correlation matrix.

Instead of the 31 or 34 industries of earlier chapters, the engineering industry of the present chapter contains 37 industries somewhat differently defined[14]. The skilled manual worker intensity is according to table 7:1 the only intensity which obtains insignificant correlation coefficients with all measures of commodity market barriers as well as K/P. Both the capital and the technical personnel intensity are significantly correlated with the average capital stock per plant, K/P, which latter variable is also correlated with the seller concentration ratio, *CONC*, and the market growth rate. The simple correlation coefficients are not so high as to necessitate the exclusion of any variable due to the risk for multicollinearity. However, the fact that K/P is significantly correlated with as many as four of the other six independent variables may cast doubt on whether significance tests of its regression coefficients are meaningful.

The second reason why the six regression coefficients do not sum up to 1.0 is that *some* of the relationships are nonlinear in terms of *some* of the independent variables. The lack of a theoretical basis for assuming given functional forms led us to investigate several alternative forms. Unlike the case in other chapters, the results in this chapter did not unambigously suggest that one form was preferable to others for all dependent and independent variables. For that reason several functional forms are presented below.

Table 7:1

A correlation matrix between measures of factor and commodity market barriers. 37 observations

	1	2	3	4	5	6	7
1 (K/L)	1.00						
2 (L_T/L)	0.02	1.00					
3 (L_Y/L_A)	0.38[a]	0.37[a]	1.00				
4 (K/P)	0.42[a]	0.47[a]	0.24	1.00			
5 *CONC*	0.17	0.28	-0.14	0.41[a]	1.00		
6 Δt	-0.18	0.15	-0.07	-0.08	0.22	1.00	
7 $\Delta C/C$	-0.35[a]	0.04	-0.14	-0.33[a]	-0.27	-0.11	1.00

[a] Significant at the 5% level (one-tail test)

K/L = horsepower/employee in 1959

L_T/L = technical personnel/employee in 1959

L_Y/L_A = the ratio of skilled manual workers to total number of manual workers in 1972 (measures in hours of work)

K/P = horsepower per plant in 1959

CONC = four firm concentration ratio multiplied by 1-(M/C) (=H/C)

Δt = decrease in tariff level of total imports 1960-1970

$\Delta C/C$ = growth rate of domestic consumption 1959-1968

7.4 *A regression analysis of the production adjustment of permanent producers*

The relationships between specialization trends of trade, $\Delta[(X-M)/C]$ and $\Delta(O/C)$, and of permanent producers on the one hand, and the explanatory variables on the other were if anything linear. Table 7:2 presents the estimated linear regression functions. The explanatory power varies from 42 to 70% and the F-values are for all three regressions significant.

Let us first compare the regression on $\Delta[(X-M)/C]$ with the regression on $\Delta[(X-M)/(X+M)]$ (cf section 7.3.2 above) as regards the regression coefficients of the factor intensities. The main difference is the capital intensity which did not significantly contribute to the

Table 7:2

Regression analysis of specialization trends, in foreign trade and for permanent firms 1959-1968. **37 engineering industries.**

Regression no.	Dependent variable	Constant	Regression coefficients (with standard errors) for *K/L*	*L_T/L*	*L_Y/L_A*	*K/P*	*CONC*	*Δt*	*ΔC/C*	R^2	F-value (degrees of freedom)
1	$\Delta\left(\frac{X-M}{C}\right)$	0.333	-0.011 (0.025)	1.684[b] (0.976)	4.594[a] (2.940)	-0.017 (0.031)	-0.322[a] (0.235)	-1.226 (2.122)	-2.119[c] (0.715)	0.419	2.981[b] (7;29)
2	$\Delta\left(\frac{O}{C}\right)$	0.694	-0.013 (0.027)	2.366[c] (1.085)	4.198 (3.270)	-0.027 (0.034)	-0.158 (0.261)	-5.874[c] (2.359)	-4.214[c] (0.795)	0.609	6.439[c] (7;29)
3	$\Delta\left(\frac{O_p}{C}\right)$	0.782	-0.018 (0.025)	2.515[c] (1.007)	3.987[a] (3.035)	-0.025 (0.032)	-0.099 (0.242)	-6.594[c] (2.190)	-4.847[c] (0.738)	0.695	9.419[c] (7;29)

[a] Significant at the 10% level (one-tail test)
[b] " " " 5% " " " "
[c] " " " 2.5% " " " "

Legend complementing those of table 7:1:

$\Delta\left(\frac{X-M}{C}\right)$ = The change in the net exports ratio in domestic consumption.

$\Delta\left(\frac{O}{C}\right)$ = The change in the total production/domestic consumption ratio; *O* is measured from plant data.

$\Delta\left(\frac{O_p}{C}\right)$ = The change in the production of permanent firms/domestic consumption ratio.

explanation of $\Delta[(X-M)/C]$. The simple correlation coefficient 0.13 suggests that this lack of significance is probably not attributable to the inclusion of other explanatory variables in table 7:2. Instead, the earlier results of chapters 4 and 6 indicate that it may be the fact that the most capital intensive industries have very small gross trade ratios $\Delta[(X+M)/C]$ which can account for the difference in results between $\Delta[(X-M)/C]$ and $\Delta[(X-M)/(X+M)]$. If this is true high transport costs for those industries constitute an entry barrier for foreign producers as well as an entry barrier for domestic producers on foreign markets.

Regression 1 of table 7:2 has two other differences compared to the results of chapter 6, one being the lack of significance for Δt and the other being the significant negative influence of *CONC*. The latter result suggests that a highly concentrated seller market in 1959 appears to have been connected with a relatively slow growth of exports and/or imports. However, that result is not obtained when instead of $\Delta[(X-M)/C]$ the dependent variable $\Delta(O/C)$ is used. The only difference between those two variables stems from the fact that O (gross output) is measured from plant data. The consequences of this are of three kinds. First, Δt becomes significant instead of *CONC*; secondly, significance is no longer obtained for L_Y/L_A; and thirdly, higher regression coefficients (in absolute terms) and significance are obtained for both L_T/L and $\Delta C/C$. The correlation coefficient between the two dependent variables of regressions 1 and 2 is 0.79.

The third regression establishes our norm of comparison for subsequent sections. Its dependent variable $\Delta(O_P/C)$ is highly correlated with both $\Delta(O/C)$ (r=0.98) and $\Delta[(X-M)/C]$ (r=0.73). The results of regression 3 are similar to those of regression 2 except for the significance obtained for L_Y/L_A. It can be noticed that no significance is registered for either *K/P* or *CONC*, variables which are usually considered as important entry barriers for new domestic firms. In a small open economy, the comparative development of permanent firms appears to be more strongly influenced by their competitiveness against foreign producers on domestic and foreign markets than by entries of domestic firms. However, the results for *K/P* are uncertain due to the many significant correlations with other independent variables. In Du Rietz [forthcoming] it is one of the most influential explanatory variables, which may be attributable to the fact that his study includes also metal industries with substantially higher K/P values than for any engineering industry.

So far we have found almost the same independent variables to be influential in explaining $\Delta(O_P/C)$ as in chapter 6 for $\Delta[(X-M)/(X+M)]$. Permanent producers have increased their specialization in human capital intensive industries with slow-growing markets and small tariff decreases. Hence, we have so far not obtained any indications of adjustment determinants which have distorted the influence of the factor intensities. Nor did we find other significant specialization determinants which may help to explain the relatively small explanatory power of the factor intensities in chapter 6. The analysis of that point is brought forward in the next section comparing the development of permanent producers with that of producers which also had production plants in 1959 but discontinued operations before 1968.

7.5 A regression analysis of adjustment through exits

Two observations must first be made. One is that the exit rates, $\Delta(O_{E1}/C)$ and $\Delta(O_{E2}/C)$, are by definition equal to the *negative* values of the 1959 production/consumption ratio of the exiting producers. Secondly, both variables appeared to be non-linearly related to L_T/L, K/P and $CONC$ and perhaps also L_Y/L_A in a way suggesting the use of semi-logarithmic functional forms or the inverse of these three variables. Both functional forms gave substantially higher explanatory values than the linear relationships found relevant for permanent firms. This departure gives additional information regarding the nature of potential exit barriers.

Regression results are presented for independent exits (regressions 1 and 2) and specialization exits (3 and 4) in table 7:3. Significance is only obtained for regressions on independent exits, the variance of which is explained to more than 50% to be compared with about 24% for specialization exits.

According to section 7.2 above the absence of factor market barriers to exit would in the medium run hypothetically imply higher exit rates for capital inten-

Table 7:3

Regression analysis of exits in 1959-1968

37 engineering industries

Regression no.	Dependent variable	Regression coefficients (with standard errors) for: Constant	K/L	L/L_T	$\log (L_T/L)$	$\log (L_Y/L_A)$	$\frac{1}{K/P}$	$\log (K/P)$	$\frac{1}{CONC}$	$\log CONC$	Δt	$\Delta C/C$	R^2 F-value (degrees of freedom)
1	$\Delta\left(\frac{O_{E1}}{C}\right)$	-0.010 (0.041)	-0.002 (0.004)	-0.001[a] (0.000)		-0.001 (0.012)	-0.005[a] (0.003)		-0.005 (0.005)		0.080 (0.300)	0.151[a] (0.107)	0.587 5.877[c] (7;29)
2		0.048 (0.046)	0.003 (0.005)		0.038[c] (0.013)	-0.016 (0.014)		-0.006 (0.011)		0.031[b] (0.017)	-0.050 (0.347)	0.015 (0.116)	0.515 4.398[c] (7;29)
3	$\Delta\left(\frac{O_{E2}}{C}\right)$	-0.086[b] (0.042)	0.000 (0.004)	0.000 (0.000)		-0.007 (0.013)	-0.000 (0.003)		-0.001 (0.005)		0.496[a] (0.312)	0.222[b] (0.112)	0.237 1.288 (7;29)
4		-0.081[b] (0.044)	0.002 (0.005)		0.002 (0.013)	-0.008 (0.014)		-0.008 (0.011)		0.010 (0.016)	0.391 (0.332)	0.199[b] (0.111)	0.245 1.342 (7;29)

[a] Significant at the 10% level (one-tail test)
[b] " " " 5% " " " "
[c] " " " 2.5% " " " "

Legend complementing those of table 7:1:

$\Delta\left(\frac{O_{E1}}{C}\right)$ = minus the ratio: production by discontinued independent firms/domestic consumption (in 1959).

$\Delta\left(\frac{O_{E2}}{C}\right)$ = minus the ratio: production by discontinued affiliates/domestic consumption (in 1959).

sive industries, i.e., the exit rates should be negatively related to the capital intensity. Moreover, the exit rates should be lower for technical personnel and skilled manual worker intensive industries, i.e., exit rates should be positively related to the two human skill intensities.

Regressions 1 and 2 give support to the assumption of such an adjustment pattern only for the technical personnel intensity. The exit rate of independent firms were lower in technical personnel intensive industries. However, in contrast to the relationship for permanent firms regressions 1 and 2 are non-linear in L_T/L (see also section 7.6 below).

As earlier mentioned the lack of significant results for the capital intensity may be attributable to the influence of high transport costs for the most capital intensive engineering products. The results of table 7:3 correspond on this point with those found for permanent firms.

The most obvious difference between the regressions on the exit rates and those on the comparative development of permanent firms is that the latter were positively related to the skilled manual worker intensity. That result may suggest the existence of exit barriers. One possibility is that large local or regional variations in the endowment of such workers combined with insufficient geographical mobility have obscured the exit patterns. If this is the case, the entry rates should at least not be positively related to the skilled manual worker intensity (cf section 7.6). In any case the lack of significance in table 7:3 gives in itself evidence of distortions which cause the adjustment of trade and production to deviate from the pattern predicted by traditional factor proportions theories[15].

As mentioned above the relatively low explanatory values of the factor intensities in chapter 6 may also be due to the rôle of barriers which are not systematically related to the factor intensities. The significance obtained in regression 1 for *K/P* and in regression 2 for *CONC* supports the existence of such barriers. Hence, independent exits were smaller the larger the average capital stock per plant and/or the larger the domestic seller concentration ratio. Both relationships are as expected.

Evidence is also obtained supporting the assumption that a rapid market growth rate means a smaller exit rate. Since this is especially the case for specialization exits, the result runs counter to the tendency for permanent firms to decrease their specialization in fast growing markets. The conclusion regarding whether or not this indicates an exit barrier of a different kind than the increasing marginal cost of capital of existing firms has to wait until we have analysed entry patterns.

There is a weak indication of that specialization exits have been larger the less the tariff decreased. If anything the opposite result would have been expected.

In summary, the approach was more successful in explaining independent exits than specialization exits. Certain results suggest the possibility of barriers to exit which may or may not have affected the trade and production adjustment caused by a possible change in Sweden's factor abundance. More firm conclusions can be drawn only after the study of entry patterns, since barriers to exit are normally also barriers to entry (cf Caves & Porter [1976]). An interesting phenomenon is the non-linear regression relationships for exit rates which contrasts to the linear regression functions found for $\Delta[(\tilde{X}-M)/C]$, $\Delta(O/C)$ as well as for permanent firms $\Delta(O_P/C)$. Because of the identical relationship between the specialization variables (cf formula (3a) the implication is that the non-linear relationships for the exit rates are counteracted by non-linear relationships for the entry rates.

7.6 A regression analysis of adjustment through entries

The entry rates are also non-linearly related to some of the independent variables used in the analysis of permanent firms. It turned out that the functional forms of the regressions obtaining the highest explanatory values were those used above for exit rates (see table 7:3). Following the two approaches of that analysis it was possible to explain 65-70% of new firm entries (regressions 1 and 2 of table 7:4) and 44-53% of diversification entries (regressions 5 and 6 of table 7:4). The high percentages of the latter entry group could, however, only be reached after the exclusion of one extreme observation[16]. Below, we shall not comment upon regressions 3 and 4, which are in-

Table 7:4

Regression analysis of entries in 1959-1968

36-37 engineering industries

Regression no.	Dependent variable	Constant	K/L	L/L_T	$\log (L_T/L)$	$\log (L_Y/L_A)$	$\frac{1}{K/P}$	$\log (K/P)$	$\frac{1}{CONC}$	$\log CONC$	Δt	$\Delta C/C$	R^2 F-value (degrees of freedom)
		Regression coefficients (with standard errors) for											
1	$\Delta\left(\frac{O_{N1}}{C}\right)$	-0.087^b (0.049)	0.007^c (0.005)	0.001^a (0.000)		0.007 (0.015)	0.003 (0.004)		0.012^c (0.005)		0.421 (0.361)	0.309^c (0.129)	0.696 9.472^c (7;29)
2		-0.167^c (0.056)	0.001 (0.006)		-0.044^c (0.016)	0.022 (0.017)		0.015 (0.013)		-0.058^c (0.020)	0.617^a (0.415)	0.464^c (0.138)	0.649 7.648^c (7;29)
3	$\Delta\left(\frac{O_{N2}}{C}\right)$	0.089^a (0.064)	-0.010^c (0.007)	0.0010^b (0.0006)		0.019 (0.019)	-0.007^a (0.005)		0.005 (0.007)		-0.014 (0.474)	-0.006 (0.169)	0.191 0.977 (7;29)
4		-0.001 (0.070)	-0.009 (0.007)		-0.035^b (0.020)	0.024 (0.021)		0.016 (0.017)		-0.013 (0.025)	0.081 (0.519)	-0.003 (0.173)	0.149 0.724 (7;29)
5		-0.020 (0.028)	0.001 (0.003)	0.0001^c (0.000)		0.002 (0.008)	-0.002 (0.002)		0.005^a (0.003)		0.103) (0.196)	0.080 (0.070)	0.528 4.475^c (7;28)
6		$-0,071^c$ (0.031)	0.001 (0.003)		-0.021^c (0.009)	0.005 (0.010)		0.007 (0.007)		-0.020^b (0.011)	0.145 (0.229)	0.107^a (0.077)	0.436 3.089^b (7;28)

[a] Significant at the 10% level (one-tail test)
[b] " " " 5% " " " "
[c] " " " 2.5% " " " "

Legend complementing those of table 7:1:

$\Delta\left(\frac{O_{N1}}{C}\right)$ = the change in the ratio: production by new, independent firms/domestic consumption (in 1968)

$\Delta\left(\frac{O_{N2}}{C}\right)$ = the change in the ratio: production by new, affiliates/domestic consumption (in 1968).

cluded in table 7:4 in order to indicate the change in results brought about by that exclusion.

The entry rates for new firms and for diversification entries were significantly larger the smaller the technical personnel intensity. The fact that both permanent firms and exits developed in contrast more favourably for technical personnel intensive production appears to indicate the existence of an entry and exit barrier associated with that production factor. A possible interpretation is that the excess demand for technical personnel or engineers combined with limited inter-firm or inter-sectoral mobility of those skills in the medium run prevented new firms from specializing in production intensive in such skills[17]. Whatever the relevant explanation is, we can draw the conclusion that adjustment through entries moved counter to the trend required by a change in Sweden's comparative advantage towards technical personnel intensive production. That finding helps explain why the trade adjustment of the 1960's was not more closely related to the explanatory power of this factor intensity. It seems safe to conclude that an earlier or more rapid increase in the supply of new engineers along with a more noticeable decrease in their relative wages should have helped to bring about a more pronounced trade adjustment of that nature.

Table 7:4 does not allow any conclusions with respect to the capital intensity for reasons discussed in commenting upon the development pattern of permanent and exiting firms. In contrast, the lack of significance for the skilled manual worker intensity in regressions on both entry and exit rates suggests that entries as well as exiting firms have not been able to draw upon the low Swedish comparative costs for such workers. According to table 7:2 only permanent firms benefited from that advantage. The larger excess demand for skilled manual workers mentioned in chapter 6 combined with low inter-regional or inter-firm mobility might have created an entry barrier at least for some firms. At any rate, the results of tables 7:3 and 7:4 indicate impediments distorting or delaying the trade adjustment that would follow upon a change towards a better comparative advantage on skilled manual worker intensive products in an economy with perfectly functioning factor markets.

According to table 7:4 a high seller concentration ratio appears to have been an effective entry barrier especially for new independent firms but also for diversification entries. In combination with the equivalentresults for the exit rates, those results may be taken to imply the existence of a major entry and exit barrier. As such it should have also affected foreign producers to export to Sweden or domestic producers to export according to regression 1 of table 7:2. Hence, it seems safe to conclude that we have found an additional reason for the limited power of the factor intensities to explain the trade adjustment patterns of the 1960's.

The capital stock per plant (K/P) receives also in table 7:4 only weak support. Whether that result is attributable to the fairly small values of (K/P) in the engineering industry or to multicollinearity between that variable and some of the other independent ones cannot be determined. Therefore, it is impossible to judge whether or not absolute capital cost advantages constitute important entry and exit barriers affecting the trade adjustment patterns.

Table 7:4 gives only weak indications that the tariff decrease established a diminishing entry barrier for new, domestic producers. The trade liberalization of the 1960's appears to have affected the production structure of permanent producers versus that of foreign ones more than entries and exits. This should not be taken to mean that the tariff decrease did not have substantial effects on entries and exits, only that the effects were not related to the size of that decrease.

The significant, positive regression coefficient for the market growth rate ($\Delta C/C$)give support to an interpretation that it is easier for both new and foreign producers to enter a market if that market grows rapidly. In addition, table 7:3 suggested that the exit rate was then smaller. Hence, there is all reason to assume that the market growth rate reflects the influence of important entry and exit barriers.

However, that barrier cannot merely be attributed to increasing marginal costs of capacity for existing producers (cf Du Rietz & Hause [1976] and Du Rietz [forthcoming]) because of the fact that exits were also correlated with the market growth rate. In our view both Swedish corporate tax laws[18] and the highly developed economy should give strong incentives for *all* domestic producers to specialize in rapidly expanding markets. The fact that it is the permanent producers which deviated from that pattern meant that Sweden's trade specialization *decreased* in fast growing engineering products due to their dominating influence as both exporters and import-competing producers. If it can be assumed that the permanent producers in the engineering industry also accounted for a large part of diversification entries and specialization exits, then we have an alternative interpretation to the increasing marginal cost of capacity hypothesis. Suppose that major, multi-plant engineering firms were already before the 1960's specialized in products with low market growth rates during that decade. Then the above mentioned incentives to adjust towards more rapidly expanding markets could be reflected in a) diversification entries and specialization exits and b) changes in intra-industry specialization. Tables 7:3 and 7:4 give evidence of that the first type of adjustment have in fact occurred. If the latter kind of adjustment is strong the analysis of Sweden's inter-commodity group specialization trend should at least reveal weaker tendencies toward decreased specialization in fast growing products than was the case above and in chapter 6.

It is important to emphasize that the reasoning on this point is based on either one of two alternative assumptions. One is that the comparative development of permanent firms (=plants) as measured by $\Delta(O_P/C)$ is due to the fact that their capital equipment is so specific that it cannot be utilized if the output mix is drastically changed towards fast growing products. That assumption is compatible with decisions which may be efficient from the point of view of the individual firm, although not for the whole economy. The alternative assumption is that the comparative development of permanent firms can be attributable to a decision-making which has been inefficient even for individual producers in the long run. A too strong emphasis may have been laid on investments in labour-saving equipment, market investments for expanding sales of old products etc. rather than investments aimed at reallocating all resources to more rapidly expanding products. This latter kind of interpretation has been increasingly emphasized in the 1970's as a main factor behind the unfavourable development of a number of large producers.

7.7 *Conclusions*

The analysis of this chapter focused on revealing the existence of certain entry and exit barriers, which might have prevented the kind of trade and production adjustment that would follow upon a change in comparative advantage of the type discussed in chapter 6. Thus, the firm adjustment patterns studied here deviated from traditional factor proportions models in not assuming perfectly functioning factor and commodity markets. While the analysis of commodity market barriers departed from theoretically and/or empirically well-known and well-established determinants, the analysis of factor market barriers could not utilize much knowledge from earlier literature. Therefore, the purpose was limited to reveal whether or not sub-groups of domestic firms differed from each other in the relationship between specialization trends and factor intensities.

The method used in studying the comparative development of permanent, new firm and diversification entries and independent and specialization exits utilized specialization measures with domestic consumption (C) instead of gross trade ($X+M$) in the denominator. Earlier chapters as well as a comparison of regressions on $\Delta[(X-M)/(X+M)]$ and $\Delta[(X-M)/C]$ indicated that the use of such measures produced unreliable results for the capital intensity. Again it appears as if high transport costs for the most capital intensive engineering industries combined to establish effective barriers against imports as well as exports.

However, the results for especially the technical personnel intensity but also for the skilled manual worker intensity indicated non-perfectly functioning factor markets in the 9-year long period. Only permanent firms increased their specialization in technical personnel and skilled manual worker intensive production. Both independent and diversification entries had a comparatively unfavourable growth in technical personnel intensive production and were in addition not at all correlated with the skilled manual worker intensity. In combination with a registered severe excess demand for both kinds of skilled personnel this result gives evidence of systematic factor market differentials of some kind in a period when Sweden appeared to have an increased comparative advantage in production intensively using those skills. In conclusion, our findings suggest some mechanisms which could have prevented or delayed the Rybczynski-type adjustment discussed in chapter 6.

From that chapter we also know about existing factor price rigidities with respect to the average wage ratios for engineers and skilled manual workers. It is true that the analysis of entry and exit barriers does not give much more information as to the nature of the factor market differentials. Thus, it is not possible to judge whether the differentials are attributable to the change in factor abundance or to more regularly predominant differentials associable with for instance the scattered population distribution in a land-abundant economy like the Swedish one, the time requirements of the educational process for the two skills investigated, differences in job careers in new versus permanent firms etc. For the purpose of the present study it is sufficient to conclude that the analysis of factor market barriers to entry and exit does not completely destroy the fundamental adjustment mechanisms of factor proportions theories. In a longer period than a decade those mechanisms might be much stronger.

Apart from factor market barriers, evidence of the influence of entry and exit barriers on commodity markets was obtained. One important barrier was a high market concentration of domestic producers in 1959 as measured by a four-firm concentration ratio multiplied by $[1-(M/C)]$ to take account of import competition. According to the results a high initial seller concentration prevented or delayed both imports and new domestic firms from entering the market.

Another variable which proved to be influential was the growth rate of domestic consumption. The comparative development of permanent firms (=permanent plants + expansion entries - contraction exits) was higher the lower that growth rate. In fact, the finding in chapter 6 of a tendency towards decreased Swedish trade specialization in fast growing markets is fully attributable to the behaviour of that group of permanent firms. New firm entries grew more and exits of independent firms relatively less in such markets. The same results were obtained for diversification entries and specialization exits, respectively.

These findings do not allow us to discriminate between alternative explanations to the failure of permanent producers to reallocate their expansion towards fast growing markets. An analysis of the relationships between specialization trends and market growth rates at lower levels of aggregation allowing other measures of specialization and market growth rates may be one way to gain more information.

In any case, the analysis of commodity market barriers has shed some light on why trade specialization trends was not more closely related to the factor intensities and the (assumed) comparative advantage of the industries. By and large factor proportions accounts of changing trade patterns have been shown to exist. They would possibly have been even more influential than they seem if either the time period covered had been longer or certain barriers to entry and exit had been made less effective.

Footnotes

1. An earlier version of this chapter was read at the Third Annual Conference on "Economics of Industrial Structure": Anti-Trust and Economic Efficiency. We are grateful to comments made by participants of that conference and acknowledge with gratitude subsequent comments from R.E. Caves.

2. The terminology of Bhagwati [1971] and Magee [1973, 1976] is in part not appropriate here since our analysis belongs to the positive branch of trade theory.

3. For references see the recent study by Magee [1976] which study delineates and analyses factor market differentials.

4. To our knowledge the analysis of this chapter is the first effort to analyse entry and exit barriers within the framework of a general equilibrium, factor proportions theory. However, as mentioned by Caves & Khalilzadeh-Shirazi [1976] several empirical studies of industrial organization have recently been recognizing the fact that most economies are open economies (cf Caves [1974], Esposito & Esposito [1971], Pagaoulatos & Sorenson [1976a and b], White [1974]). The limited purpose of our own analysis and its rather narrow interest on barriers affecting trade adjustment under an assumed change in Sweden's factor abundance prevented us from directly benefiting from those earlier studies. The underlying theory of this chapter is even more tentative than in other chapters of the monograph although we have tried to keep the analytical framework of chapter 6 as much as possible.

5. This latter group of entries is comparatively small, explaining why it was not separated from the first group.

6. Which should not cause too much trouble *if* in fact inter-industry differences in transport costs etc. explain very much of the gross trade ratio $(X+M)/C$ and *if* those differences remained stable between 1959 and 1968.

7. Before 1968 Sweden did not have any substantial direct subsidies to the industrial sector. The exception to this rule are the regional industrial incentives mainly in the form of investment subsidies introduced in the middle of the 1960's

for the Northern part of Sweden. However, that location aid did not reach sufficient size within the period to have had any substantial structural impact even for the region. Policy-imposed entry and exit barriers of possible influence and of a more indirect nature were for instance certain capital market regulations, the dimensioning of the educational system especially for engineers and manual worker skills and the regulation of the yields of private educational investment through scholarships, and the tax system. Although the impact of such policies cannot be ruled out as unimportant, we had no opportunity to estimate directly that impact in the present study. However their influence cannot be controlled and separated from other factors directly studied in this chapter.

8. The reader is referred to Du Rietz [1975] and [forthcoming] for a study of entry barriers of that kind.

9. Klevmarken [1974] gives evidence of a short run sensitivity of the engineering salary movements to the excess demand for engineers.

10. According to the accounts of his theory published in for instance Grubel [1967, 1970], Hufbauer [1970] and Wells [1972].

11. In part this result may be attributable to the fact that both heterogeneity measures were significantly, negatively correlated with a variable measuring seller concentration (see below). The relative standard deviation of the low price was thus negatively correlated with the adjusted four-firm concentration ratio (r=0.44) indicating that for technologically relatively homogeneous industries it was easier to maintain a strong seller concentration.

12. The slight change in period from 1960-1970 to 1959-1968 did not very much affect the regression coefficients for the factor intensities as can be seen from the following regression (standard errors within parentheses; a = significance at the 5% level and b = significance at the 2.5% level, one-tail tests):

$$\Delta\left(\frac{X-M}{X+M}\right) = 0.086 - \underset{(0.019)}{0.058^{b}}(K/L) + \underset{(0.707)}{1.331^{a}}(L_T/L) + \underset{(2.485)}{4.775^{a}}(L_Y/L_A).$$

$$R^2 = 0.346$$
$$F(3;33)=5.831^{b}$$

A comparison with regression 4 of table 6:4 shows that the above regression obtains stronger significance and a higher explanatory power than the corresponding regression for the period 1960-1970. It is difficult though to judge whether this can be attributable to the change in period or to, for instance, the increased number of and somewhat changed industries.

13. Cartels or collusion between Swedish and foreign producers affecting the Swedish market must be assumed rare or at least randomly distributed. No information could be obtained regarding such agreements.

14. On this point cf Du Rietz [forthcoming].

15. With perfectly functioning factor and commodity markets. This conclusion is not only reached from table 7:3 but relies also on the non-significant correlations between the two exit rates and the skilled manual worker intensity. The lack of significance in table 7:3 for that intensity does not appear to be due to multicollinearity.

16. The excluded industry is the marine engine manufacturing (SNI 38413), which had one extremely large diversification entry during the period.

17. Gort [1962] found in contrast that diversification entries in US industry 1929-1954 were larger in technical personnel intensive, rapidly expanding industries.

18. Especially the depreciation allowances which are very liberal in an international comparison.

Chapter 8

THE COMMODITY GROUP SPECIALIZATION PATTERN OF 15 INDUSTRIAL COUNTRIES IN 1970

8.1 Outline of the analytical problems

Chapters 8 and 9 have three purposes in common. First, they aim at investigating to what extent the inter-industry results of chapters 4 and 6, respectively carry over to Sweden's commodity group specialization. A second complementary aim is to find out whether inter- and intra-commodity group specialization is systematically related to the technology characteristics of the products. Finally, both chapters try to relate Sweden's trade specialization to that of other industrial countries. Chapter 9 goes on to answer whether the negative specialization trend found for Sweden's net export ratio in chapter 6 holds at lower levels of aggregation, for other measures of specialization, for other industrial countries and for a somewhat different time period.

The abandonment of the industry analysis has several disadvantages, all related to the fact that there are no available data on the factor use of commodity groups. These disadvantages are described in section 8.2[1]. On the other hand the methodology used has the advantage of allowing the simultaneous analysis of inter- and intra-commodity group specialization. In addition, there are several other advantages associated with the commodity group analysis. One is the possibility of including a third specialization measure, namely the world export share (X_S/X_{OECD}) discussed in chapter 2 and another that other possible trade determinants can be investigated. A third advantage is that the observations are more homogeneous, technologically and otherwise, and that the variance among the observations increases in going from about 30 industries to more than one hundred commodity groups. One indication of the larger homogeneity is the increased standard deviation of the net export ratio from 0.30 at the industry level to 0.49 at the commodity group level[2].

8.2 Methodology

There are several methodological issues necessary to discuss in the following. One is the rôle of the technology proxy variable in investigating the inter-commodity group specialization within a factor proportions framework. Another issue is how to measure the intra-commodity group specialization and the varying heterogeneity of commodity groups. A third issue is whether or not it is reasonable to assume irreversible factor intensities across countries save for possible differences caused by intra-industry or intra-commodity group specialization.

8.2.1 The technology proxy and its utilization

Chapter 5 presented the ton price as a possible proxy variable for inter-industry and inter-commodity group differences in capital and technical personnel intensities[3]. Appendix B discusses that variable in more depth.

The methodology of chapters 8 and 9 is based on the strategic assumption that ton price differences of the integrated processes reflect differences in primarily capital and technical personnel intensities of that part of the processes which is value added of the engineering industry (as shown in appendix B). Those differences are called technological to emphasize that we compare different production functions.

Recall that in such a comparison a higher capital intensity means a lower ton price while a higher technical personnel intensity is associated with a higher ton price. Hence, within the engineering sector it is true that products which require more human capital obtain a higher ton price while those characterized by a high degree of mechanization receive a lower price per ton. Two concepts are introduced to distinguish between products with different technologies. An engineering product directly utilizing much *human* but little *non-human* capital per employee will be labelled a technologically sophisticated or complex product. A product with the opposite characteristics will be called a technologically standardized or simple product[4].

The estimated ton price relationships of chapter 5 and appendix B are based on Swedish data. However, the analysis of chapters 8 and 9 involves also other industrial countries. Instead of using the ton price of Swedish trade it would ideally be better to use the corresponding price of world trade[5]. The acceptance of the latter ton price depends on the validity of two strategic assumptions. One is irreversible factor intensities for homogeneous products and the other is equalized export and import prices for such products. The former assumption implies that the relative requirements of raw materials and intermediate inputs per ton of output are also irreversible and the latter that the prices of such raw materials and inputs are the same internationally.

In conclusion, the methodology of chapters 8 and 9 does not require more restrictive assumptions than those of earlier chapters. However, it should be observed that non-equalized factor prices mean that the *size* of the regression coefficients of the factor intensities in regressions on the ton prices of production (or of exports and imports) differs. As shown in appendix B such differences were obtained from regressions on Swedish exports and imports[6].

The conclusion is that it should be possible to use ton prices of world trade. Lacking such prices, the best approximation found was ton prices of exports from OECD-Europe[7]. The simple correlation coefficient between this export ton price and the Swedish export ton price at the level of 106 commodity groups was 0.69[8].

Disaggregation from an industry to a commodity group level increases the variation in technological differences, as indicated by the increased range in ton prices from $1,000-$34,000 at the industry level to $200-$59,400 at the commodity group level. As shown by appendix table C:4 the latter level includes in particular more observations with low ton prices[9]. Hence, the disaggregation might widen the differences in capital intensity, which did not reach very high values at the industry level compared to other parts of the manufacturing industry.

This concludes the discussion of the properties of the ton price variable as such. Next we have to find out whether it is a useful variable in analysing the factor proportions theory.

The answer depends crucially on two subquestions: 1) Is the country abundant in human capital or non-human capital or both?, and 2) Are engineering technologies such as to produce a systematic relationship between technical personnel and capital intensities?

According to chapters 4 and 5 all indications were that Sweden in 1970 was not specialized in industries intensive in technical personnel. The conclusion was not as clear for capital intensive products since the home market share was larger the more capital intensive the industry. The discussion of the impact of tradability led to the finding that *part* of this relationship appeared to be due to the low tradability of the most capital intensive industries. That result together with the fact that those industries did not even have as high capital intensities as the average value of the whole manufacturing industry suggests that Sweden should rather be assumed to be abundant in non-human capital than in human capital. However, the results derived are not at all unanimous and it is only the need of a specified hypothesis that persuades us to draw such a conclusion.

The second subquestion relates to how the technology varies among engineering products. Are technical personnel intensive products generally also capital intensive or on the contrary capital extensive or is there no systematic relationship at all? Looking back at table 4:2 we can notice that few industries are intensive in both kinds of capital. The most obvious exception to this is manufacture of engines and turbines ranking as the 9th most capital intensive and the 3rd most technical personnel intensive industry. There is a weak negative correlation between the two intensities (-0.22), which would have been stronger with an exclusion of that industry and the pulp and paper mill machinery industry (with ranks 8 and 6).

If anything our conclusion should be that technical personnel intensive engineering products are generally not capital intensive and vice versa. Thus, a country which is abundantly supplied with technical personnel should be specialized in products with high ton prices and another one with an abundance in physical capital should be specialized in engineering products with low ton prices. According to our earlier results Sweden should rather be expected to have the latter pattern than the former one.

This finding is startling because it is contrary to both what is generally expected for a highly developed economy and to the particular hypothesis of the product cycle theory (cf Hirsch [1967] and related neotechnology theories). Since the ton price appears to have properties which fit in very well with the product cycle theory (if we disregard the tricky issue whether technological sophistication has anything to do with the age of engineering products), we have good opportunities to test the explanatory power of that theory. In fact, given the lack of clarity with respect to Sweden's factor abundance situation and the availability of only one technology variable, the prospects of testing that theory appear to be better on the commodity group level than the prospects of testing the factor proportions theory. At best the results for Sweden of chapters 8 and 9 can be used as complements to the findings of earlier chapters in our investigations of factor proportions explanations to the Swedish trade pattern.

8.2.2 *Measures of intra-commodity group specialization and technological heterogeneity*

The disaggregation to the commodity group level does not secure homogeneous observations even with respect to technological characteristics. The use of the ton price of OECD-Europe (P_{jE}) as our technology proxy for commodity groups makes it possible to study explicitly also the specialization within commodity groups for Sweden (and other industrial countries). Suppose that perfectly identical products receive the same world market price (P_i) and that there are no differences in weights per unit of such products. Then the ton price ratio of a commodity group j, i.e. P_{jS}/P_{jE}, where P_{jS} = the Swedish export ton price of that group, can be used to measure how Sweden's intra-commodity group specialization differs from the pattern of OECD-Europe, since for a given commodity group:

$$\frac{P_{jS}}{P_{jE}} = \frac{\sum_i P_i \frac{q_{iS}}{q_S}}{\sum_i P_i \frac{q_{iE}}{q_E}}; \qquad i = 1,\ldots,n \text{ of commodity group } j$$

and P_i = ton price of a homogeneous commodity i

q_{iS} = Swedish export volume (in metric tons) of that commodity

q_S = total Swedish export volume (in metric tons) of all commodities $i = 1,\ldots,n$

q_{iE} = export volume (in metric tons) from OECD-Europe of commodity i

q_E = total export volume (in metric tons) of commodities $i = 1,\ldots,n$ from OECD-Europe.

According to our assumptions it is the relative distribution of the export volume of Sweden compared to that of OECD-Europe, which creates ton price ratios deviating from 1.0. For technologically homogeneous commodity groups, P_{jS}/P_{jE} is a measure of Sweden's product differentiation in products which are close substitutes but with varying qualities. Appendix table C:4 presents P_{jS}/P_{jE} for the 106 commodity groups. It ranges from 0.15 up to 9.22 indicating large differences in intra-commodity group specialization.

However, the commodity groups are far from technologically homogeneous. Even worse from the point of view of analysing determinants of the specialization patterns, the degree of heterogeneity varies substantially among the commodity groups. Chapter 5 presented possible measures of the degree of heterogeneity for industries. Those measures relied on the possibility of using ton prices of commodity groups *within* each industry. Since the 106 commodity groups establish the lowest possible level of OECD statistics for the 1960's similar heterogeneity measures could not be used for commodity groups.

However, an alternative method of measuring heterogeneity differences of commodity groups is to use the export ton prices of the European countries. A heterogeneity measure using these ton prices relies on the assumption that the eleven countries have specialized *within* commodity groups to approximately the same extent for all commodity groups. This tends to be the case if the factor proportions theory is relevant at the micro level, if the countries differ much with respect to their factor abundance and if the homogeneous products differ with respect to factor intensities.

The heterogeneity measure is defined as the coefficient of variation of the export ton price of European countries. Appendix table C:4 shows its values for the 106 commodity groups. It varies from 0.16, i.e., 16% of the average (unweighted) ton price of the countries, up to 1.92. The corresponding coefficient of variation estimated for the whole engineering sector utilizing ton prices at the level of the 106 commodity groups is 1.61. Although more homogeneous observations have been obtained at the commodity group level compared to the industry level, there are obviously still some heterogeneous observations.

The purpose of this monograph does not demand a detailed study of intra-commodity group patterns of specialization. However, it requires an investigation into the possibility of systematic *national* differences in those patterns and especially how Sweden differs in this respect from other industrial countries. Indirectly this analysis is begun in the next section, which considers the question: whether or not the technology can be assumed to be internationally the same for the commodity groups.

8.2.3 Are technology differences internationally similar?

Earlier chapters have reported two findings suggesting a substantial intertemporal stability in technological differences. One finding was that the relative differences in capital and technical personnel intensities of 33 industries remained stable over a 15-year period. The other finding was a strong correspondence between the 1964 and 1970 export ton prices of OECD-Europe at the 106 commodity groups level. That finding may be seen as a substantial stability both in the underlying relative capital and technical personnel intensities and in the corresponding factor price ratios[10].

Our next query is whether there is evidence of a similar technological stability across countries in a given year, i.e., 1970. According to chapter 3, the industrial countries differ substantially in their factor price ratios, in particular as regards the relative wages of skilled workers. Combined with non-neoclassical production functions such differences may generate factor intensity reversals.

Unable to compare directly factor intensities at the industry level, we shall make use of the information hidden in the export ton prices of the about one hundred commodity groups of nine European countries[11]. We have already pointed out the inter-country differences in export ton prices for individual commodity groups. A regression analysis between the ton prices of a given country ($P_{j,k}$) and the ton price of OECD-Europe ($P_{j,E}$) will provide information on whether or not such deviations are rare and possibly due to odd intra-commodity group specialization in certain commodity groups. The explanatory power of the regression will be low if there are many commodity groups which are characterized by such a specialization. It seems likely that lower R^2 values will be obtained for small countries than for large ones. High R^2 values can be interpreted to mean that a) not many commodity groups have a strong intra-commodity group specialization (causing $P_{j,k}$ to deviate substantially from $P_{j,E}$) and b) further support is obtained for the assumption that either production functions can be approximated by neoclassical production functions or the inter-country differences in factor price ratios are too small to create factor reversals.

Table 8:1 provides regression results for the nine countries. Each regression as well as the coefficient for P_E is highly significant. Five of the regressions produced R^2 values ranging from 0.67 to 0.91 with lower explanatory powers for four small countries.

The results of table 8:1 do not give much support to an assumption that there are *major* inter-country differences in the factor requirements of homogeneous products. The results appear instead to emphasize the possible importance of intra-commodity group (or industry) specialization in causing $P_{j,k}$ to deviate from $P_{j,E}$. This conclusion agrees also with the fact that engineering industries and commodity groups are technologically heterogeneous.

Table 8:1 appears to demonstrate important inter-country differences in intra-commodity group specialization. Belgium and Switzerland are the extreme countries in this respect. Belgium had according to regression 1 specialized in sophisticated products *within* technologically unsophisticated commodity groups while Switzerland had specialized more in such products *within* the technologically sophisticated commodity groups. Germany was specialized in much the same way as Switzerland. Denmark had a similar pattern to Belgium. The regression relationships of Sweden and Austria did not deviate much from the characteristics of a 45°-line through the origin. Italy differed from the other countries with regression coefficients lower than 1.0 in having also a low intercept term.

In summary, the estimated relationships do not have properties which seem to be damaging for an assumption of non-reversible factor intensities of (homogeneous) engineering products. However, they do appear to emphasize the existence of important national differences in intra-commodity group specialization. Therefore, we shall try to establish a better empirical base for those differences before studying inter-commodity group specialization patterns.

8.3 The intra-commodity group patterns of specialization

Two kinds of systematic inter-country differences in intra-commodity specialization patterns shall be investigated. Both may be attributable to the technology of the products and the factor abundance of the countries. However, it was impossible to construct a sufficient theoretical and empirical basis for establish-

Table 8:1

The regression relationships between the export ton price of a country and that of OECD-Europe in 1970

Regression no.	Country	Regression line (with standard error)	R^2 F-value (degrees of freedom)
1	Belgium-Luxemburg	$P_B = 2.27 + 0.60^* P_E$ (0.10)	$R^2 = 0.25$ F(1;102) = 33.79*
2	The Netherlands	$P_N = -0.49 + 1.16^* P_E$ (0.04)	$R^2 = 0.88$ F(1;101) = 753.65*
3	Germany	$P_G = -0.44 + 1.30^* P_E$ (0.04)	$R^2 = 0.91$ F(1;103) = 1 056.25*
4	France	$P_F = 0.26 + 1.03^* P_E$ (0.07)	$R^2 = 0.68$ F(1;104) = 216.54*
5	Italy	$P_I = 0.90 + 0.72^* P_E$ (0.05)	$R^2 = 0.67$ F(1;102) = 209.24*
6	Denmark	$P_D = 2.17 + 0.66^* P_E$ (0.09)	$R^2 = 0.38$ F(1;99) = 60.74*
7	Sweden	$P_S = 1.96 + 0.91^* P_E$ (0.09)	$R^2 = 0.48$ F(1;104) = 94.86*
8	Austria	$P_A - 0.92 + 1.03^* P_E$ (0.11)	$R^2 = 0.48$ F(1;97) = 90.82*
9	Switzerland	$P_{SZ} = 0.29 + 1.52^* P_E$ (0.08)	$R^2 = 0.77$ F(1;102) = 340.39*

* significant at a lower level than 1%.

ing testable hypothesis derived from the factor proportions theory.

First, we shall investigate which countries tend to be specialized in sophisticated products and which in contrast are specialized systematically in a more standardized assortment. Table 8:2 gives the proportion of commodity groups with ton price ratios (P_k/P_E) larger than 1.0 for nine industrial countries. Three countries deviate from the others in having a relatively high proportion of ton price ratios above 1.0:

Table 8:2

Proportion of commodity groups with export ton price ratios above 1.0 in nine industrial countries 1970

Country	Proportion of commodity groups with $P_k/P_E > 1.0$ in per cent	Number of commodity groups included
Belgium-Luxemburg	31	104
The Netherlands	48	103
West Germany	75	106
France	36	106
Italy	35	105
Denmark	54	101
Austria	42	99
Switzerland	83	101

Source: OECD, Commodity Trade Statistics, Series C, Exports 1970.

Switzerland and West Germany and also Sweden[12]. Belgium in particular but even France and Italy appeared in contrast to have high proportions of ton price ratios lower than 1.0. Although the British ton price ratio could be calculated for only 66 commodity groups, Great Britain may be expected to belong to this latter group of countries with only 38% of the commodity groups having ratios above 1.0.

The findings reported in table 8:2 seem to be generally in accordance with commonly held views about differences in the development levels between the two groups of countries. If that is the case the results may also be associable with the abundance of capital and skilled labour, although there are certainly other possible explanations to these findings.

The second kind of systematic inter-country differences in intra-commodity group specialization is whether the ton price ratios of the countries are systematically related to the technological sophistication of the commodity groups. That seemed to be the case according to the regressions presented in table 8:1. However, the extent of intra-commodity group specialization may also be related to the heterogeneity of the commodity groups. The regressions of table 8:1 did not take account of tha factor.

Therefore, we shall by regression analysis relate P_k/P_E to both the heterogeneity measure, henceforth labelled h, and the export ton price of OECD-Europe, P_E. It is expected that each country will show a positive relationship between P_k/P_E and h.

The regression results are presented in table 8:3. Six of the regressions receive significant F-values and seven show positive regression coefficients for h. Hence, almost every country had a larger P_k/P_E the more heterogeneous the commodity group was according to our measure h. Four countries have in addition significant negative regression coefficients for P_E, namely Italy, Sweden, Denmark and Switzerland. Those relatively small countries have thus specialized in relatively sophisticated products in commodity groups which are capital intensive and/or technical personnel extensive. It should be emphasized that those conclusions differ from those found in section 8.2.3 in connection with table 8:1. That difference is attributable to the influence of the heterogeneity measure in the regressions of table 8:3, which is very powerful in explaining the Swiss ton price ratio[13]. None of the countries have systematically specialized in products with relatively low ton prices in heterogeneous commodity groups.

Although it is difficult to explain the patterns of intra-commodity group specialization we can conclude that there are substantial differences among the countries studied. The Swedish pattern was characterized by a specialization in sophisticated products in about 60% of the commodity groups and there was a tendency for the ton price ratios to be higher or the product differentiation on high quality products to be stronger the less sophisticated the commodity group was. Having found systematic inter-country differences in intra-commodity group specialization, we shall next study the inter-commodity specialization patterns to see whether there are similar differences among countries on that level. That study tries to take account of the influence of the former specialization on the latter one.

8.4 The inter-commodity group pattern of specialization

The ensuing analysis includes 15 industrial countries, which were member countries of the OECD in 1970. As is evident from table 8:4 they differ much from each other with respect to size of engineering exports. West Germany was the largest exporter of engineering products,

Table 8:3

Regressions on the ton price ratios in 1970 exports of nine industrial countries

Regression no.	Country	Constant	Regression coefficients (with standard errors) for h	P_E	R^2	F-value (degrees of freedom)
1	Belgium-Luxemburg	0.561[c]	1.315 (0.444)	-0.009 (0.016)	0.081	4.448[c] (2;101)
2	The Netherlands	1.032	0.129 (0.223)	-0.002 (0.008)	0.004	0.184 (2;100)
3	West Germany	1.025	0.276[c] (0.112)	0.004 (0.004)	0.065	3.593[b] (2;103)
4	France	0.915	0.153[a] (0.118)	0.005 (0.004)	0.032	1.695 (2;103)
5	Italy	0.963	0.088 (0.087)	-0.012[c] (0.003)	0.145	8.664[c] (2;102)
6	Sweden	0.520	2.303[c] (0.420)	-0.020[a] (0.015)	0.232	15.532[c] (2;103)
7	Denmark	0.818	1.329[c] (0.402)	-0.020[b] (0.012)	0.115	6.376[c] (2;98)
8	Austria	0.833	0.660[a] (0.447)	0.003 (0.015)	0.023	1.136 (2;96)
9	Switzerland	-1.147	8.259[c] (0.681)	-0.068[c] (0.024)	0.601	75.900[c] (2;101)

[a] Significant at the 10% level (one-tail test)

[b] " " " 5% " " " "

[c] " " " 2.5% " " " "

Note: The differences in number of observations are due to the fact that for some countries there were no available export quantities in metric tons published for a few commodity groups.

Table 8:4

Trade with engineering products of 15 industrial countries in 1970

Country	% of total OECD trade		% of total trade of the country		Net export ratio in %
	Exports	Imports	Exports	Imports	
Canada	6.5	12.0	31	52	-15
USA	20.7	20.9	38	30	15
Japan	9.9	3.4	39	10	60
Belgium-Luxemburg	3.4	5.5	23	28	- 9
The Netherlands	2.9	6.5	19	28	-24
West Germany	22.2	10.3	51	20	50
France	7.7	8.9	34	27	8
Italy	6.8	5.5	40	21	26
Great Britain	10.4	6.3	42	17	39
Norway	0.4	1.7	12	26	-53
Sweden	3.4	3.9	39	32	8
Denmark	1.1	2.2	25	29	-21
Austria	1.0	2.2	29	35	-21
Switzerland	2.4	3.4	38	30	- 1
Finland	0.4	1.6	12	34	-52
Sum of above	98.8	94.1	37	26	18
Total OECD	100.0	100.0	37	26	15

Source: OECD, Commodity Trade Statistics, Series C, 1970.

earning about half its total export income on such products. Its net export ratio was 50% due to relatively small imports. The US was the second largest exporter but also a large importer of engineering goods.

Great Britain and Japan each accounted for about 10% of total OECD exports. Both countries also had large net export ratios. These four countries accounted for about 63% of total OECD exports but only 41% of its total imports. All small countries except Sweden were net importers of engineering goods.

The following regression analysis of the commodity

group specialization pattern is divided into three subsections. The first treats the four large countries and the second all other countries (except Sweden). Results are reported for two specialization measures, the net export ratio and the world export share. The Swedish results, which are presented in the third subsection, include in addition a third measure, namely the home market share, and also a few more independent variables. The regression analysis for Sweden deviates also in another respect. The more solid empirical basis on Sweden's comparative advantage derived in chapters 3-5 made it possible to establish testable hypotheses regarding the signs of the regression coefficients.

8.4.1 The specialization pattern of the US, Japan, West Germany and Great Britain

Three and alternatively four explanatory variables are tested against the two specialization measures. First and foremost is P_E, our technology proxy, about whose sign we have no *a priori* hypothesis. Another independent variable is the size of the world market, measured by total OECD exports, X_W. Our hypothesis for that variable is that large countries tend to specialize in products with a large world market. That hypothesis rests on an underlying assumption, namely that such products are characterized by relatively large economies of scale. The proximity of a large domestic market combined with scale economies is assumed to establish a comparative advantage in those products.

However, the variable X_W measures OECD (or world) exports for commodity groups rather than for (homogeneous) products. Controlling, hopefully, the influence of varying heterogeneity with h, we shall nevertheless assume X_W to have a more positive influence on the trade specialization of a large country than of a small one.

Our heterogeneity measure is thus also included as an independent variable. Small countries are presumed to have easier access to the world market for heterogeneous commodity groups with a relatively small world market. In addition to h, the ton price ratio P_k/P_E is also included for the above nine countries with data on P_k. No hypothesis regarding its possible impact on the inter-commodity group specialization pattern is specified.

Some of the regression results are presented in table 8:5. Except for one case, all regressions produce very low R^2 values as was also the case for the Swedish inter-industry specialization in chapters 4 and 5. Several regressions of table 8:5 are, despite the low R^2 values, significant. The approach was most successful in explaining the British net export ratio, for which 18% of the variance was explained.

The most unambigous results were obtained for Great Britain. Both the net export ratio and the world export shares were significantly negatively related to the ton price, P_E. Hence, Great Britain tended to be specialized in unsophisticated engineering products. That conclusion held not only at the commodity group level but also for the intra-commodity group specialization.

For the remaining three large exporters, the regression analysis revealed that the two specialization measures were not correlated significantly with the same explanatory variables. Thus, while the US net export ratio was larger for more homogeneous commodity groups, the world export share was uncorrelated with the heterogeneity measure. Instead that share was higher for commodity groups with a large world market and a high ton price.

Similarly, the Japanese net export ratio was negatively related to the ton price, while the world export share was positively related to the same variable. That tendency of a less negative (more positive) relationship between the world export share and the ton price than between the net export ratio and the ton price holds also for West Germany. In fact, we shall see that the same tendency is obtained for almost all countries. Since all fifteen countries had a non-positive correlation between the net export ratios and the ton price, it seems as if all countries tend both to export and import sophisticated products and that only some of them have an export profile sufficiently strong in such products as to generate significant ton price coefficients in regressions for world export shares. According to table 8:5 only the US and Japan had such a profile among the four large countries.

As is obvious from appendix table C:4 there are more commodity groups with low ton prices than with high ones. Plots have revealed that most countries specialized greatly in at least a few of the former ones while almost none had very high net export ratios or world market shares for any of the latter ones.

Table 8:5

Regressions on the net export ratios and the world export shares of the U.S., Japan, West Germany and Great Britain in 1970

Re-gres-sion no.	Country and dependent variable	Constant	Regression coefficients (with standard errors) for X_W	P_E	h	P_k/P_E	R^2	F-value (degrees of freedom)
	USA							
1	$\frac{X-M}{X+M}$	0.4554	-0.0027 (0.0043)	-0.0069 (0.0063)	-0.4429[c] (0.1855)		0.066	2.386[a] (3;102)
2	$\frac{X}{X_W}$	0.1305	0.0021[c] (0.0009)	0.0018[a] (0.0014)	0.0201 (0.0396)		0.066	2.382[a] (3;102)
	Japan							
3	$\frac{X-M}{X+M}$	0.6290	-0.0030 (0.0036)	-0.0107[c] (0.0054)	0.0628 (0.1567)		0.045	1.614 (3;102)
4	$\frac{X}{X_W}$	0.1121	-0.0012 (0.0010)	0.0034[c] (0.0015)	0.0094 (0.0443)		0.064	2.304[a] (3;102)
	West Germany							
5	$\frac{X-M}{X+M}$	0.6086	-0.0009 (0.0020)	-0.0061[c] (0.0030)	-0.0785 (0.0884)	-0.0353 (0.0738)	0.056	1.490 (4;101)
6	$\frac{X}{X_W}$	0.2337	-0.0002 (0.0008)	0.0003 (0.0012)	-0.0512[a] (0.0347)	0.0093 (0.0290)	0.022	0.561 (4;101)
	Great Britain							
7	$\frac{X-M}{X+M}$	0.4737	-0.0001 (0.0028)	-0.0192[c] (0.0041)	0.0304 (0.1209)		0.175	7.215[c] (3;102)
8	$\frac{X}{X_W}$	0.1234	-0.0003 (0.0005)	-0.0016[c] (0.0008)	-0.0069 (0.0228)		0.042	1.472 (3;102)

[a] Significant at the 10% level (one-tail test)
[b] " " " 5% " " " "
[c] " " " 2.5% " " "

Legend:

$\frac{X-M}{X+M}$ = net export ratio in 1970

$\frac{X}{X_W}$ = world export share in 1970

X_W = Total OECD-exports 1970

P_E = metric ton price of exports of OECD-Europe in 1970

h = degree of heterogeneity in 1970

P_k/P_E = the 1970 ratio between the export ton price of country k and that of OECD-Europe.

In conclusion, our regression approach was not very successful in explaining the commodity group specialization of the four large countries in 1970. There was no general tendency for large countries to specialize in homogeneous commodity groups with large world markets. However, the finding that Great Britain specialized in unsophisticated and the US and Japan in sophisticated products appears to be in accordance with the conventional wisdom.

8.4.2 The specialization pattern of ten other OECD-countries

The regression results for the ten medium-sized or small OECD-countries (excluding Sweden) are reported in table 8:6. At least one of the two regressions are significant for most of the countries. The explanatory values for the significant regressions varied from about 10% (Canada, Norway, Denmark and Austria) up to almost 40% (Belgium-Luxemburg). Thus, evaluated in this way, the success of our approach varies a great deal. As explained in section 8.2 this may very well be due to the fact that only if a country has a strong comparative advantage in *either* human capital *or* non-human capital intensive products, there will be a close relationship between the specialization measures and the ton price variable.

Canada and Belgium specialized in 1970 in commodity groups with large world markets (X_W) while Norway, Finland and Austria avoided specializing in such commodity groups. It appears as if the size of the world market is not uniformly important for the specialization of small countries.

Three small countries hold relatively strong positions in simple or unsophisticated products. They were Belgium, Norway and Finland. For Belgium, at least, the same conclusion held also with respect to the intra-commodity group specialization. Hence, Great Britain and Belgium had similar trade patterns in 1970 as regards the technology characteristics of the products. The Belgian specialization was also pronounced in heterogeneous commodity groups. Accordingly, Belgium was specialized in heterogeneous, technologically unsophisticated commodity groups with a large world market. It also had a stronger product differentiation on high quality products within those groups. It appears as if the theory formulated by the Belgian economist J. Drèze does receive support in table 8:6 as regards

Table 8:6

Regressions on the net export ratios and world market shares of ten other OECD-countries in 1970

Regression no.	Country and dependent variable	Constant	X_W	P_E	h	P_K/P_E	R^2	F-value (degrees of freedom)
	Regression coefficients (with standard errors) for							
	Canada							
1	$\frac{X-M}{X+M}$	-0.5054	0.0074[c] (0.0037)	0.0017 (0.0055)	-0.0811 (0.1610)		0.047	1.665 (3;102)
2	$\frac{X}{X_W}$	0.0275	0.0015[c] (0.0004)	-0.0001 (0.0006)	-0.0156 (0.0182)		0.129	5.020[c] (3;102)
	Belgium-Luxemburg							
3	$\frac{X-M}{X+M}$	-0.4164	0.0060[b] (0.0031)	-0.0147[c] (0.0046)	0.4606[c] (0.1399)	0.0256 (0.0293)	0.204	6.353[c] (4;99)
4	$\frac{X}{X_W}$	-0.0083	0.0005 (0.0004)	-0.0017[c] (0.0006)	0.1389[c] (0.0187)	-0.0100 (0.0039)	0.387	15.630[c] (4;99)
	The Netherlands							
5	$\frac{X-M}{X+M}$	-0.2512	-0.0028 (0.0029)	0.0018 (0.0043)	0.0739 (0.1267)	-0.0310 (0.0555)	0.021	0.522 (4;98)
6	$\frac{X}{X_W}$	0.0395	-0.0004 (0.0003)	0.0002 (0.0004)	0.0093 (0.0118)	-0.0041 (0.0052)	0.037	0.937 (4;98)
	France							
7	$\frac{X-M}{X+M}$	-0.0824	0.0022 (0.0034)	-0.0054 (0.0051)	0.2083[a] (0.1483)	0.0346 (0.1299)	0.032	0.831 (4;101)
8	$\frac{X}{X_W}$	0.0715	-0.0003 (0.0006)	0.0001 (0.0009)	-0.0058 (0.0250)	0.0142 (0.0203)	0.007	0.170 (4;101)
	Italy							
9	$\frac{X-M}{X+M}$	0.6677	-0.0025 (0.0034)	-0.0189[c] (0.0055)	-0.3702[c] (0.1500)	-0.1594 (0.1664)	0.163	4.871[c] (4;100)
10	$\frac{X}{X_W}$	0.1117	-0.0005 (0.0006)	-0.0008) (0.0009)	-0.0480[b] (0.0248)	-0.0080 (0.0275)	0.049	1.284 (4;100)
	Norway							
11	$\frac{X-M}{X+M}$	-0.3928	-0.0030 (0.0029)	-0.0065[a] (0.0043)	-0.3180[b] (0.1245)		0.085	3.147[b] (3;102)
12	$\frac{X}{X_W}$	0.0089	-0.0001[b] (0.0001)	-0.0002[b] (0.0001)	-0.0047[b] (0.0025)		0.081	2.975[b] (3;102)

cont.

Regression no.	Country and dependent variable	Constant	Regression coefficients (with standard errors) for X_W	P_E	h	P_K/P_E	R^2	F-value (degrees of freedom)
	Denmark							
13	$\frac{X-M}{X+M}$	0.0017	-0.0038 (0.0036)	-0.0117[c] (0.0054)	-0.4159[c] (0.1942)	-0.0511 (0.0453)	0.123	3.374[c] (4;96)
14	$\frac{X}{X_W}$	0.0316	-0.0004 (0.0004)	-0.0005 (0.0005)	-0.0181 (0.0194)	-0.0009 (0.0045)	0.030	0.753 (4;96)
	Finland							
15	$\frac{X-M}{X+M}$	-0.2225	-0.0069[b] (0.0038)	-0.0162[c] (0.0056)	-0.3591[c] (0.1644)		0.130	5.097[c] (3;102)
16	$\frac{X}{X_W}$	0.0070	-0.0001 (0.0001)	-0.0002[b] (0.0001)	-0.0030 (0.0027)		0.048	1.719 (3;102)
	Austria							
17	$\frac{X-M}{X+M}$	-0.1547	-0.0066[b] (0.0036)	-0.0077[a] (0.0055)	0.3088[b] (0.1674)	-0.0349 (0.0369)	0.104	2.723[b] (4;94)
18	$\frac{X}{X_W}$	0.0260	-0.0004[a] (0.0003)	-0.0003 (0.0004)	-0.0051 (0.0132)	-0.0003 (0.0003)	0.028	0.665 (4;94)
	Switzerland							
19	$\frac{X-M}{X+M}$	-0.1414	-0.0050 (0.0043)	-0.0010 (0.0066)	0.1985 (0.2869)	-0.0187 (0.0263)	0.023	0.571 (4;99)
20	$\frac{X}{X_W}$	-0.0232	-0.0001 (0.0002)	0.0001 (0.0004)	0.0193 (0.0161)	-0.0017 (0.0015	0.022	0.561 (4;99)

[a] Significant at the 10% level (one-tail test)
[b] " " " 5% " " " "
[c] " " " 2.5% " " " "

Legend: See table 8:5

Note: As shown by the varying degrees of freedom the number of observations is sometimes less than 106. This is due to the fact that ton prices of exports from certain countries were unavailable for a few commodity groups. Regressions were also run including all 106 observations but excluding the ton price ratio for those countries. As far as signs and significance of the remaining independent variables the results of those regressions were the same as those reported above.

Belgium (cf the statements of that theory in Hufbauer [1970] and Grubel [1967, 1970]).

The net export ratios of Italy, Denmark and Austria were also higher the smaller the ton price. Italy, Norway, Denmark and Finland were specialized in relatively homogeneous commodity groups while at the same time Belgium, France and Austria had the opposite pattern. There is obviously no general tendency for small countries to be specialized in heterogeneous commodity groups with small world markets.

In conclusion, for most of the fifteen industrial countries only low explanatory values were obtained although many regressions and regression coefficients became significant. Belgium and to some extent also Great Britain and Italy were exceptions in obtaining also relatively high explanatory values. Thus, the 1970 patterns of commodity group specialization were as difficult to explain as the Swedish pattern of inter-industry specialization.

8.4.3 The Swedish commodity group specialization

As mentioned above, the analysis of the commodity group specialization of Sweden can utilize more dependent and independent variables than that for the 15 other industrial countries. Apart from the net export ratio and the world market share, it is possible to investigate the Swedish home market share. The regression analysis of the latter share uses C, domestic consumption, instead of X_W, OECD exports, as a market size measure. In addition, the influence of the tariff level and of varying tradability is tested. The hypothesis regarding each independent variable is specified in turn.

Recall first the arguments of section 8.2 above, based on the results of chapters 4 and 5 that Sweden ought to be abundant in capital rather than in technical personnel. The null-hypotheses are therefore that the three specialization measures are non-negatively correlated with the ton price. Each regression of table 8:7 gives support to a rejection of that hypothesis. The Swedish engineering industry was in 1970 specialized in unsophisticated, i.e. capital intensive/technical personnel extensive commodity groups. A comparison with tables 8:5 and 8:6 indicates in fact that Sweden can be grouped together with Great Britain, Belgium, Norway and Finland all of which countries had negative regression coefficients for the ton price in

Table 8:7

Regressions on the net export ratio, the world market share and the home market share of Sweden in 1970

Re-gres-sion no.	Depen-dent vari-able	Con-stant	Regression coefficients (with standard errors) for							R^2	F-value (degrees of freedom)
			X_W	C	P_E	h	P_S/P_E	t	r		
1	$\frac{X-M}{X+M}$	0.0401			-0.0147[c] (0.0056)					0.063	6.952[c] (1;104)
2	X/X_W	0.0362			-0.00052[a] (0.00035)					0.020	2.124 (1;104)
3	H/C	0.6010			-0.0162[c] (0.0032)					0.199	25.654[c] (1;103)
4	$\frac{X-M}{X+M}$	0.1414	0.0013 (0.0038)		-0.0143[c] (0.0056)	-0.2104 (0.1857)	-0.0114 (0.0374)			0.087	2.393 (4;101)
5	X/X_W	0.0395	-0.0000 (0.0002)		-0.0005[a] (0.0004)	-0.0040 (0.0120)	-0.0009 (0.0024)			0.024	0.630 (4;101)
6*	H/C	0.6138		0.0000[c] (0.0000)	-0.0153[c] (0.0032)	0.1572[a] (0.1038)	-0.0308[a] (0.0210	-2.2858[b] (1.1829)		0.266	7.192[c] (5;99)
7*	H/C	1.4444		0.0000 (0.0000)	-0.0107[c] (0.0029)	0.0733 (0.0894)	-0.0191 (0.0180)	0.1601 (1.0818)	-1.1010[c] (0.1775)	0.473	14.673[c] (6;98)

[a] Significant at the 10% level (one-tail-test)
[b] " " " 5% " " " "
[c] " " " 2.5% " " " "

Legend complementing those of table 8:5:

H/C = home market share in 1970

P_S/P_E = the 1970 ratio between the export ton price of Sweden and that of OECD-Europe.

t = tariff level on total Swedish imports measured as the ratio between tariff revenues and total imports

r = a measure of the trade exposure of commodities measured by $(X_W-C)/(X_W+C)$.

* These regressions include only 105 observations. No reasonable value was obtained for SITC 7325 regarding Swedish gross output since the published figure was lower than Swedish exports to the extent that not only the Swedish sales to the domestic market became negative but also domestic consumption.

regressions on both the net export ratios and the world market shares. Sweden deviates from both Great Britain and Belgium in that the intra-commodity group specialization tended to lie in products with a relatively high ton price. Both Sweden and Belgium had higher ton price ratios (P_k/P_E) the lower the ton price (P_E) was.

Table 8:7 gives further indications of the tendency of the net export ratio to be more negatively correlated with the ton price than the world export share. However, it is obvious that the Swedish home market share is even more negatively related to the ton price. According to regression 3 that variable alone explains almost 20% of the variance of the home market share, a finding comparable with the results of chapter 4 which showed a close correspondence between home market share and both the capital intensity (positive sign) and the technical personnel intensity (negative sign). To some extent these relationships are attributable to the influence of a varying tradability and to the possible correlation between tradability and factor intensities. It was judged worthwhile to investigate at the commodity group level.

Regressions 6 and 7 provide further information on that point. Notice first that in regression 6 all five independent variables become significant. Thus, the home market share was larger the larger the domestic consumption and the heterogeneity of the commodity groups. According to our small vs large country hypothesis in section 8.4.1 the latter result should be expected but the former runs counter to the same hypothesis. The home market share was smaller in 1970 the higher the Swedish ton price ratio (P_S/P_E) and the higher the tariff level. The latter, somewhat confusing result was also obtained at the industry level. If anything, the weak negative sign of P_S/P_E should be taken to indicate that for Sweden the intra-commodity group specialization in high quality products appears to be a way of avoiding competitive pressures on the assortment within each group with relatively large world (and probably also domestic) markets. That interpretation is consistent with the earlier finding of negative relationships between on the one hand P_S/P_E and P_E and on the other the three specialization measures and P_E. Sweden has specialized in relatively unsophisticated commodity groups but has avoided the most common assortment within those groups by specializing in high quality or odd products.

However, the results of regression 6 are substantially altered if a new independent variable is included (cf regression 7). That variable is a measure of the varying trade exposure of the products. It is defined as $(X_W - C)/(X_W + C)$ and labelled r. Recalling the assumption of identical demand functions in Sweden and the rest of the world, C approximates a proportion of total world consumption, C_W, which is a constant over the 105 commodity groups. Thus r should be highly correlated with $(X_W - C_W)/(X_W + C_W)$ which varies from -1.0 for non-traded goods up to 0.0 when all consumption is imports.

The properties of the variable r have been earlier discussed in Ohlsson [1973] chapter 3 and shall not be discussed extensively here. Suffice it to mention which commodity groups have the lowest and highest r values (cf appendix table C:4). Low r values and thus limited trade exposure were obtained for iron and steel constructions, tins, cans and metal boxes, metal containers for transport or storage and safes. According to earlier chapters fabricated metal products of this kind have relatively high transport costs and should therefore have a comparatively low tradability. Commodity groups with high r values are knives, scissors, forks and spoons, pins, instruments, etc., all of which appear to be easily tradable over large distances.

Regression 7 in table 8:7 shows that the variable r is highly influential in explaining H/C. The fact that it reduces the coefficient for P_E (cf regression 6) is compatible with the results of the residual analysis of chapter 4. The conclusion of that chapter and now also of the present one is that the home market share appears to be somewhat positively related to the capital intensity and negatively to the technical personnel intensity even after taking account of the impact of tradability. The positive correlation between the ton price (P_E) and r[14] is significant (cf table 8:8) but not so large as to completely remove the influence of P_E on trade specialization.

In conclusion, the results of this section strengthen the conclusions of chapters 4 and 5 that Sweden was not specialized in technical personnel intensive prod-

Table 8:8

Correlation matrix between dependent and independent variables of the regressions in table 8:7

	$\frac{X}{X_W}$	$\frac{X-M}{X+M}$	H/C	X_W	P_E	h	P_S/P_E	C	t
$\frac{X-M}{X+M}$	0.63[a]								
H/C	0.28[a]	0.51[a]							
X_W	0.01	0.07	-**						
P_E	-0.14	-0.25[a]	-0.45[a]	-0.04					
h	-0.06	-0.16	0.02	-0.23[a]	0.06				
P_S/P_E	-0.05	-0.07	-0.05	-0.07	0.09	0.47[a]			
C*	0.07	0.08	0.19[a]	-**	-0.14	-0.25[a]	-0.12		
t*	-0.04	-0.10	-0.15	-**	0.10	-0.05	-0.04	0.29[a]	
r*	-0.22[a]	-0.12	-0.62[a]	-**	0.29[a]	-0.06	0.02	-0.11	0.32

[a] Significant at the 5% level (one-tail test).

* For reasons given in table 8:7 the number of observations is 105 instead of 106 for these three rows.

** Not estimated due to the difference in number of observations.

Legend: See tables 8:5 and 8:7.

ucts in 1970. If anything, it was instead specialized in products with low requirements of technical personnel and high capital intensities. Chapters 4 and 5 indicated also a weak, but significant specialization in skilled manual worker intensive products.

The inclusion of the r variable in regression 7 meant also that four of the five independent variables in regression 6 became insignificant. That result is obtained although the correlation coefficients between r and those four variables are not very high. It is difficult to arrive at a satisfactory explanation for this change in results.

8.5 Conclusions

One of the purposes of chapter 8 was to find out whether the conclusions of chapter 4 (and 5) were specific to the industry level of aggregation and to the specialization measures used in those chapters. The study of the Swedish pattern of commodity group specialization provides evidence of that this was not the case. First of all, the technology of the products explain rather little of Sweden's net export ratio or world market share. They are more successful in explaining the home market shares at both levels of aggregation.

Secondly, Sweden was in 1970 specialized in technologically simple rather than complex engineering products, i.e. in capital intensive/technical personnel extensive products rather than the reverse. The fact that the results on this point held for all three measures of specialization at the commodity group level may be attributable to the increased variance in capital intensities obtained by disaggregation. At the industry level the highest capital intensity was not even as high as that of the whole manufacturing industry. Hence, the results of chapter 8 may indicate that Sweden may be specialized in capital intensive products in its total manufacturing trade.

Another purpose of the chapter was to provide a comparison between Sweden and each one of 14 industrial countries with regard to specialization. That comparison led to the conclusion that Sweden belonged to the group of countries which specialized in unsophisticated engineering goods. However, Sweden differed from those countries in its intra-commodity group specialization, differentiating more systematically towards products of high qualities or relatively high sophistication. That differentiation was stronger the more unsophisticated the technology of the commodity group as a whole. Thus, the intra-commodity group specialization pattern but not the inter-industry or inter-commodity group pattern appears to be more consistent with the common belief that a highly developed, small country like Sweden should be specialized in sophisticated products.

With few exceptions the regression analysis of trade specialization of the 14 countries did not generate high explanatory values despite of the many significant F-values. The Belgium specialization pattern was best explained ($R^2 = 0.39$).

With regard to the regression coefficients of the ton price variable, the fifteen countries can be grouped in four categories. First, there are those countries which uniformly specialized in unsophisticated products according to both specialization measures. That group included Sweden, Great Britain, Belgium, Norway and Finland.

A second group of countries are those which had a significant negative correlation only between the net export ratio and the ton price. West Germany, Italy, Denmark and Austria belong to that group. A third group of countries had specialization patterns which were uncorrelated with the ton price and a fourth group consisted of those whose world export shares were positively related to the ton price. Canada, the Netherlands, France and Switzerland are classified in the third group while the USA and Japan belong to the fourth.

Looking upon the ton price variable as a type of product cycle measure for the sophistication of products, the results obtained for these four groups of countries should be related to the development level of the countries. However, it is difficult to find any systematic pattern in this respect. Japan should not be expected to be classified together with the USA. Sweden ought to belong to the third or fourth group rather than the group with the most distinguished specialization in unsophisticated products.

Even with respect to the size of the domestic market the results fail to show a systematic pattern. It is neither true that countries with large domestic markets had specialized in homogeneous commodity groups with large world markets nor that small countries avoided specializing in such commodity groups.

In conclusion, the elements of trade theories other than the factor proportions theory, which have been tested in this chapter, also fail to provide powerful explanations. It appears as if no single theory can offer more than a minor contribution to explaining any country's trade specialization in a given year. It remains to be seen whether it is easier to explain changes in trade specialization for the industrial countries as was the case for Sweden at the industry level.

Footnotes

1. That methodology has earlier been described in Ohlsson [1973] and [1974].
2. Cf the use of the net export ratio for the study of two-way trade in for instance Grubel & Lloyd [1975] and Ohlsson [1975].
3. The ton price differences in general reflect those two factor intensities rather than being limited to hold at the inter-industry level. Of course, individual observations may be odd and deviate from this normal case at lower aggregation levels. However, that risk was controlled to some extent by plotting the relationships in order to see whether in fact it is odd or extreme observations that determine the sign and size of the estimated regression coefficients.
4. Sometimes the concept *technically* sophisticated processes is used for a technique which is more non-human capital intensive than another one. That comparison involves *either* a plant at two different points in time with different factor price ratios and/or with technical improvement between those points *or* different plants with different factor price ratios or different capital vintages. That terminology is associable with a given product or process while our use of the concept *technologically* sophisticated underlines that we compare different products or processes.
5. Since the Swedish ton price of a given commodity group may not be representative of the output mix of that group in world trade but rather of an odd intra-commodity group specialization. On this point reference is made to chapter 5 which is focused on that particular aggregation problem.
6. Those differences may simply be interpreted to mean that the shadow price of a given factor at world market prices differs from going factor prices in each given country unless factor prices are equalized internationally. At going differences in factor prices between countries the factor intensities are different which may explain the above observed differences.
7. No data were available for Canada, the U.S. and Japan in OECD statistics at the detailed commodity group level used here. In addition, for certain commodity groups there was no British or no Norwegian export quantity published in metric tons.
8. It would have been much higher had two commodity groups been excluded, namely nuclear reactors (SITC 7117) and electronic tubes, transistors, etc. (SITC 7293). The Swedish export ton price of those two groups was almost ten times as large as

that of OECD-Europe, suggesting a strong intra-commodity group specialization in odd products with high ton prices.

9. Only one commodity group receives a substantially larger ton price than the highest industry ton price.

10. Cf however, the discussion of the concluding section of chapter 6 on this point.

11. As mentioned above Great Britain and Norway lacked the relevant data for a number of the commodity groups.

12. At least for Switzerland and West Germany it can be rejected (at the 5% level) that this frequency is randomly determined if the observations can be assumed to be binomially distributed around a median value of 1.0.

13. The correlation coefficient between h and P_E is 0.06 (cf table 8:8 below). Thus, multicollinearity has not caused the difference in results between tables 8:1 and 8:3.

14. Lipsey & Weiss [1974] found that the estimated ocean transport rate was higher the higher the ton price of the product. Similar conclusions are drawn in Norström [1974]. He shows, however, that products with high ton prices despite higher transport costs can carry the higher transport costs of truck, train and airflight transportation compared to sea transportation.

Chapter 9

TRENDS IN COMMODITY GROUP SPECIALIZATION OF 14 INDUSTRIAL COUNTRIES 1964-1970

9.1 Outline of the analytical problems

The purpose of the present chapter is to investigate whether the findings of chapter 6 hold more generally. One task is to see whether the development of Sweden's inter-industry specialization can be extended to lower levels of aggregation. Another task is to analyse the specialization trends of other industrial countries in search of systematic changes in their comparative advantage. The ton price methodology described in chapter 8 allows in addition a further widening of the theoretical framework to include, for instance, trade determinants proposed by the product cycle theory. However, it was impossible to construct formal tests which would discriminate unambigously between that theory and the factor proportions theory. As mentioned in chapter 8 the unavailability of factor intensities at the commodity group level means that the methodology is better suited for application of the former theory than of the latter one.

The need to have a uniform data classification system for all countries and for the whole period limited the sample to 1964-1970. The shorter time period together with the changes in independent variables from the industry analysis restrict the comparability of the results of the industry and commodity group analysis.

The following section discusses the development of the specialization in total engineering trade of 14 industrial countries. Finland could not be included here because of the fact that no data were reported for that country in OECD statistics of 1964. Section 9.3 searches for negative specialization trends of the type found at the industry level for Sweden (see chapter 6). That analysis tests whether that trend is attributable to a particular type of intra-industry specialization. A deeper look into this issue is taken in section 9.4, which studies the intra-commodity group specialization. The last analytical section, 9.5, seeks explanations for changes in inter-commodity group specialization of the 14 countries.

9.2 Changes in total engineering specialization of industrial countries

Chapter 8 gave some figures on the relative size of total engineering exports for each one of the 15 industrial countries and on whether or not they were large net exporters of such products. Table 9:1 presents figures on the development between 1964 and 1970 of the aggregate world export shares and net export ratios of these countries (excluding Finland). In 1964 the USA, West Germany and Great Britain had particularly high net export ratios and world export shares. Each one of those countries experienced a decline in both measures by 1970. Almost all small coun-

Table 9:1

Aggregate net export ratios and world market shares of industrial countries' engineering trade

	Net export ratio		World export share	
Country	1964	1970	1964	1970
Canada	-56.0	-14.6	2.3	6.5
USA	55.7	15.1	25.9	20.8
Japan	43.2	59.5	5.3	9.9
Belgium-Luxemburg	-17.5	- 8.9	3.0	3.4
The Netherlands	-25.2	-24.2	3.1	2.9
West Germany	63.8	49.6	23.8	22.3
France	5.1	8.4	6.7	7.7
Italy	14.4	25.7	5.5	6.8
Great Britain	54.2	38.6	15.4	10.4
Norway	-63.3	-53.4	0.3	0.4
Sweden	1.0	8.4	3.4	3.4
Denmark	-27.0	-21.4	1.1	1.1
Austria	-26.1	-21.0	1.0	1.0
Switzerland	- 3.7	- 0.6	2.7	2.5
Sum of above countries	26.1	18.6	99.6	98.8
Total OECD	22.4	15.4	100.0	100.0

Source: OECD, Commodity Trade Statistics, Series C, Exports. Imports 1964 and 1970.

tries had negative net export ratios in 1964. Their ratios increased substantially by 1970. For some small countries the world export ratio also increased.

Table 9:1 gives the impression that countries which were specialized in 1964 in engineering products decreased their specialization on such products during the period 1964-1970 with the opposite development occurring for countries which were initially specialized in other products. The only exception to this rule is Japan. Its rapid economic growth and the consequent upgrading of its development level may account for an increasing net export ratio and world market share.

The observed negative inter-country relationship in table 9:1 between changes in the aggregate measures and initial levels in 1964 can be compared with the negative inter-industry specialization trend for Sweden (cf figure 6:1). Before trying to interpret the results there is reason to investigate whether the latter type of negative specialization trends hold at the commodity group level for Sweden and the 13 other industrial countries.

9.3 The existence of negative specialization trends at the commodity group level

Recall from chapter 6 that a negative specialization trend was found to exist only for the Swedish net export ratio. The development of the home market share was not at all related to the 1960 home market shares and it was impossible to study the world market share. The analysis of chapter 6 is complemented here by investigating how changes in all the three measures relate to the initial level at the 106 commodity groups level[1]. Only the home market share is studied for the whole decade with results as follows:

$$\Delta(H/C)_{60-70} = \underset{}{0.033} - \underset{(0.066)}{0.242^{a}} \qquad R^2 = 0.115 \quad F(1;103) = 13.323^{a}$$

[a] Significant at the 1% level (one-tail test).

In contrast to results of the industry level, a negative specialization trend exists at the commodity group level for the home market share. Since the intra-industry specialization explanation discussed in chapter 6 suggests that a negative trend at one level of aggregation tends to disappear at lower levels, the above regression result is yet another indication for a rejection of that explanation.

Table 9:2 presents 28 regression relationships of the specialization trends for the 14 industrial countries. Obviously, negative trends are obtained for both Swedish net exports ratio and the world export share despite the shorter time period studied. Again, all indications suggest other explanations than one based purely on the intra-industry trade argument.

The relationships between changes in specialization 1964-1970 and the 1964 level of specialization were all linear[2] despite the boundedness of both dependent and independent variables. As earlier mentioned this was true for almost all relationships for which the boundedness of dependent variables inclined us to expect non-linear relationships. The fact that the relationships are linear (if at all existing) simplifies the regression analysis in what follows.

Table 9:2 and figures 9:1 and 9:2 give the overwhelming impression that all countries have had negative specialization trends for 1964-1970. This is true for all 14 countries with respect to their net export ratios and for 11 in terms of their world export shares. Should those trends be interpreted to mean important changes in the international division of labour? No definite answer can be given to this question in the present study because of two things. First, our study is restricted to cover only engineering trade. Secondly, as shown in figures 9:1 and 9:2 no slope coefficient is lower than -1.0. Thus, none of the countries has reversed its pattern of engineering trade specialization between 1964 and 1970. Only an extension of the period could possibly prove that shifts in trade specialization have in fact occurred. However, a recent study (see Caves & Uekusa [1976]) demonstrates that Japan has in fact reversed its trade specialization since the second world war. For other countries changes in factor endowments may or may not have led to a complete reversal of the trade pattern.

The analysis on this point will therefore be restricted to answering two questions, namely a) Is the specialization trend and the 1964 specialization pattern related to the factor requirements of the products with opposite signs? and b) Is it possible to derive indications as to which countries might have led the suggested development?

The first question will be analysed in sections 9.4 and 9.5 below. The discussion of the second question can start with a reference to table 9:1. If one or a

Figure 9:1

Regression lines for 14 industrial countries between changes in net export ratios 1960-1970 and the 1964 net export ratio

106 commodity groups

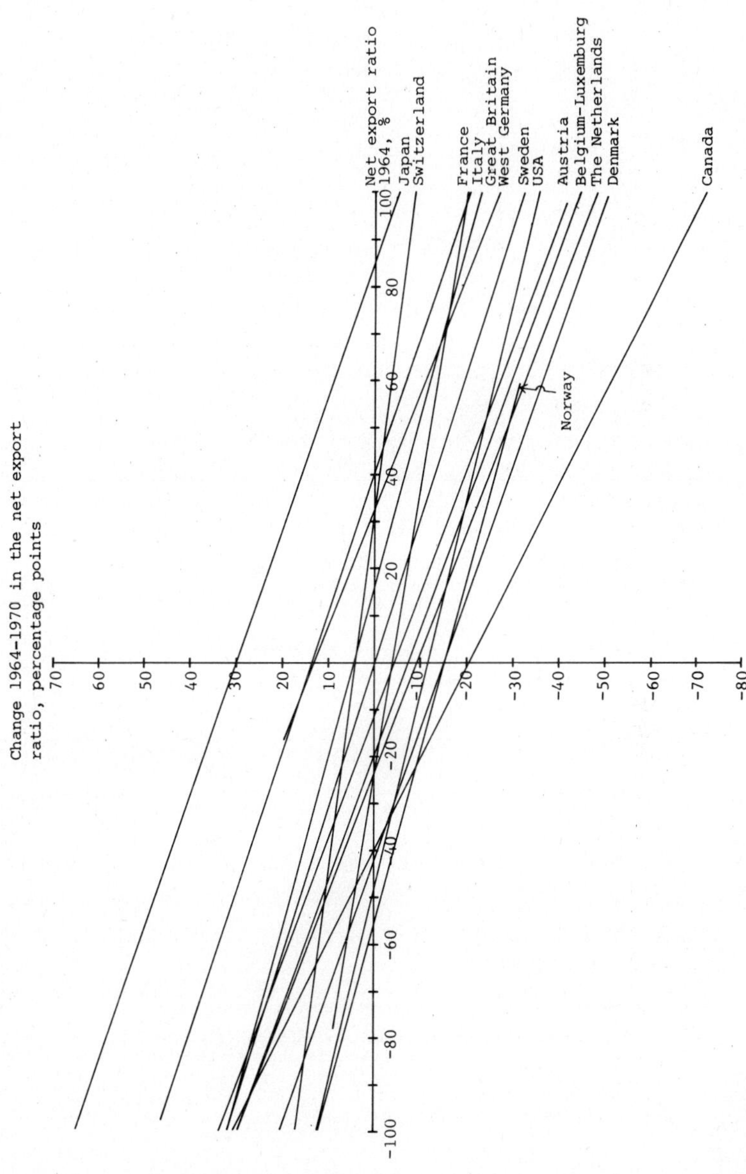

Note: The regression results are presented in table 9:2.

Figure 9:2

Regression lines for 13 industrial countries[a] between changes in world export shares 1964-1970 and the 1964 world export shares

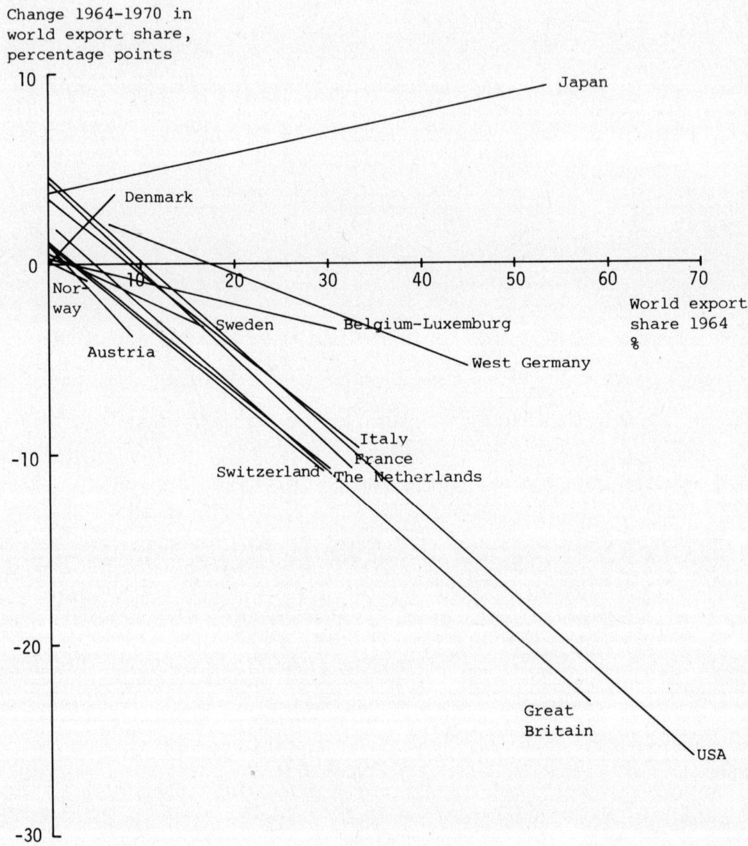

[a] The regression for Canada was insignificant.

Note: The axes have different scales because of the need to separate the regression lines of the figure. The regression results are presented in table 9:2.

Table 9:2

Regression between changes in specialization 1964-1970 and the 1964 specialization for 14 industrial countries

106 commodity groups

	Net export ratio			World export share		
Country	Con-stant	Regression coefficient (standard error)	R^2 F-value (degrees of freedom)	Con-stant	Regression coefficient (standard error)	R^2 F-value (degrees of freedom)
Canada	-0.201	-0.523[b] (0.088)	0.254 35.333[b] (1;104)	0.014	0.116 (0.203)	0.003 0.325 (1;104)
USA	-0.116	-0.243[b] (0.060)	0.137 16.461[b] (1;104)	0.046	-0.434[b] (0.055)	0.379 63.481[b] (1;104)
Japan	0.291	-0.363[b] (0.050)	0.336 52.718[b] (1;104)	0.036	0.103[a] (0.066)	0.023 2.491[a] (1;104)
Belgium-Luxemburg	-0.073	-0.379[b] (0.068)	0.229 30.956[b] (1;104)	0.003	-0.119[b] (0.030)	0.133 15.934[b] (1;104)
The Netherlands	-0.099	-0.404[b] (0.066)	0.266 37.763[b] (1;104)	0.014	-0.394[b] (0.045)	0.420 75.314[b] (1;104)
West Germany	0.130	-0.415[b] (0.081)	0.201 26.207[b] (1;104)	0.034	-0.190[b] (0.063)	0.079 8.949[b] (1;104)
France	-0.042	-0.162[b] (0.068)	0.053 5.779[b] (1;104)	0.044	-0.464[b] (0.144)	0.091 10.388[b] (1;104)
Italy	0.124	-0.346[b] (0.078)	0.160 19.743[b] (1;104)	0.035	-0.392[b] (0.093)	0.147 17.858[b] (1;104)
Great Britain	0.043	-0.280[b] (0.066)	0.148 18.072[b] (1;104)	0.024	-0.431[b] (0.040)	0.528 116.146[b] (1;104)
Norway	-0.152	-0.277[b] (0.068)	0.139 16.795[b] (1;104)	0.002	-0.269[b] (0.069)	0.127 15.156[b] (1;104)

cont.

Country	Net export ratio			World export share		
	Constant	Regression coefficient (standard error)	R^2 F-value (degrees of freedom)	Constant	Regression coefficient (standard error)	R^2 F-value (degrees of freedom)
Sweden	-0.005	-0.329[b] (0.064)	0.202 26.359[b] (1;104)	0.008	-0.247[b] (0.049)	0.199 25.809[b] (1;104)
Denmark	-0.159	-0.377[b] (0.066)	0.236 32.100[b] (1;104)	-0.003	0.547[b] (0.238)	0.048 5.257[b] (1;104)
Austria	-0.049	-0.392[b] (0.070)	0.234 31.686[b] (1;104)	0.010	-0.519[b] (0.202)	0.060 6.596[b] (1;104)
Switzerland	0.040	-0.031[b] (0.041)	0.091 10.456[b] (1;104)	0.010	-0.395[b] (0.044)	0.438 80.971[b] (1;104)

[a] Significant at the 5% level (one-tail test).
[b] " " " 2.5% " " " "

few countries start a development towards a major change in the patterns of international trade they must be large. According to table 9:1 at most four countries are sufficiently large, namely West Germany, the U.S., Great Britain and Japan. The U.S. and Great Britain are the big losers in terms of world export shares while Japan increased its share of OECD exports substantially.

Table 9:2 shows that the negative trend in both the US and the British world export shares were strong ones in terms of slope coefficients and R^2 values. Japan, on the other hand, had a positive trend for its world export share despite the fact that a strong negative trend was registered for its net exports ratio. West Germany showed negative trends for both measures.

In conclusion, enough support is obtained for a hypothesis that it may be changes in the comparative position of a few large countries which account for the negative specialization trends. The strength of those trends varies substantially among the smaller countries. Sweden had according to table 9:2 neither particularly strong nor weak trends.

Table 9:6

The percentage proportion of commodity groups with ton price ratios above 1.0 in nine industrial countries 1964 and 1970

	1964		1970	
Country	Proportion in %	Number of commodity groups	Proportion in %	Number of commodity groups
Belgium	33	105	31	104
The Netherlands	41	103	48	103
Germany	62	106	75	106
France	54	106	36	106
Italy	45	106	35	105
Denmark	58	101	54	101
Sweden	64	106	60	106
Austria	46	98	42	99
Switzerland	86	100	83	101

Note: Similar calculations for Great Britain gave the result for 1964 37%(of 70 commodity groups) and for 1970 38 % (of 66 commodity groups).

merely be interpreted as yet another consequence of the strong forces working towards a possible reversal of certain countries' trade patterns.

Similar to the results of the foregoing chapter, the analysis of the present one points at inter-country differences in intra-commodity group specialization. Sweden deviated from other countries only in its tendency to specialize in products with high ton prices within many commodity groups— perhaps a high quality, product differentiation? However, there were only very weak indications of a diminished comparative advantage in capital intensive/technical personnel extensive production. In some respects, the Swedish intra-commodity group specialization developed more as the Belgian or Italian one than as such small advanced countries as the Netherlands and Switzerland.

9.5 *Trends in inter-commodity group specialization patterns*

9.5.1 *Outline of the methodology*

The main purpose of the following regression analysis is to seek for explanations of the negative specialization trends discussed above. Recall that a factor proportions account was fairly successful in explaining the Swedish development at the industry level. Therefore it appears reasonable to follow the approach of chapter 6 as much as possible also for other *small* economies. However, as demonstrated in section 9.2 both West Germany and the U.S. accounted for as much as 20% of total OECD exports. Although questionable, the same approach is pursued for these two countries by treating them as small economies. In the two country framework, the "large country" is therefore the rest of the world which accounted for a world export share four times as high as either one of those two countries.

Instead of domestic factor intensities the technology differences are measured by the proxy variable P_{E64}. That variable was negatively correlated with K/L and positively with L_T/L (and with L_Y/L_A). Hence, only two possible general changes in comparative advantage can be detected, namely a change from a capital abundance towards a technical personnel abundance or the opposite development. For obvious reasons, the testing of the factor proportions accounts in explaining the negative specialization trends is not as promising at the commodity group level as at the industry level, especially since the period is rather short.

However, an independent variable can be utilized at the commodity group level which was not available at the industry level, namely ΔP_E. The concluding section of chapter 6 reported results which indicated larger ton price increases in 1964-1970 for commodity groups with high ton prices in 1964. Two alternative interpretations of this tendency were offered: either a) a relative price increase occurred in technical personnel intensive/capital extensive products or b) the technical personnel intensity increased more and/or the capital intensity less for those products. Regardless of interpretation a small country with an increased abundance in technical personnel and a decreased one in capital should be expected to have an additional incentive to increase its specialization in technologically sophisticated products.

The third variable suggested by the factor proportions approach of chapter 6 was the market growth rate. For Sweden the growth rate of domestic consumption was found to be very influential for the changing specialization. However, the analysis of the present chapter cannot utilize that variable for other industrial countries. Instead, the growth rate of total OECD exports, $\Delta X_W / X_{W64}$ will be used. The earlier assumption of identical demand functions between the small country and the rest of the world is obviously no longer sufficient. In addition, we shall have to assume that the growth rate of world trade will approximate the growth rate of domestic consumption of each country fairly well.

So far we have merely discussed the factor proportions approach in general terms. With the exception of Sweden, no specific hypothesis can be formulated regarding the signs of the regression of the above three independent variables. Instead, we shall restrict the analysis on this point to whether there are factor proportions explanations to the observed negative specialization trends. For this to be the case it has to be demonstrated that the change in specialization, ΔS , and the 1964 specialization, S_{64}, receive opposite signs of the regression coefficient for P_{E64} . Due to inherent methodological difficulties discussed above the absence of such a result does, however, not rule out a factor proportions account of that trend.

There are, of course, other possible explanations to that trend suggested by other trade theories. In the following, product cycle accounts will be tested along with factor proportions accounts. All three independent variables of above might under certain assumptions have been motivated also within a product cycle framework. Therefore, it will be difficult to discriminate between the two theories unless the latter theory specifies the signs of the regression coefficients or can be represented by additional independent variables.

The original product cycle theories by Hirsch [1967] and Vernon [1966] offer two different hypotheses regarding the specialization of industrial countries. The Hirsch version proposes that small advanced countries like the Netherlands, Sweden and Switzerland should be specialized in new products and that the larger industrial countries should be specialized in products in the growth phase of their product life cycle. Vernon [1966] specifies instead that large, advanced countries like the USA had a comparative advantage in new products.

In conclusion, it is not possible to arrive at unambigous hypotheses regarding the sign of the regression coefficients for each country. The product cycle theory is not very well defined in terms of which national characteristics will provide the differences in comparative advantage in products for different phases of their life cycle. Likewise it is not very well developed as regards the operationalization of the characteristics of the "product life". For a cross-sectional analysis the age of the products is impossible as an independent variable for obvious theoretical and empirical reasons. Therefore, we have to define empirically the various technology and market characteristics of the life cycle of the products.

The more than 100 commodity groups are assumed to approximate product groups in different life cycles. Some of the groups include products which are not very close substitutes in terms of technology or market characteristics compared to other groups, but the main differences in this respect are assumed attributable to the differences in product life.

None of the countries is a developing country and most of the commodity groups consist of products which were not produced very much in developing countries. Therefore our coverage does not provide enough variation to constitute a fullfledged test of the product cycle theory. That limitation has to be remembered in the following.

The market growth rate, here measured by $\Delta X_W/X_{W64}$, is one of the product life characteristics. It is usually assumed that over the life cycle products grow faster on the market the younger they are. It is not true, however, that the growth rate of products in a certain period of time is as closely related to the actual age of the product. Instead, the opposite direction of causality must usually be assumed, namely that products with rapid market growth have for that reason not reached a stage when competition has led to standardized technologies and to similar qualitative properties of the product variants. As will be discussed below our methodology includes other independent variables which measure the differences in standardization. For that reason the market growth rate can be assumed to reveal merely if a country adjusts its trade specialization easily to fast- or slow-growing products.

Four variables are used to capture differences in the level and change in technology and in the degree of standardization of the product groups. P_{E64} and ΔP_E have already been discussed above. The heterogeneity degree, h_{64}, is utilized as a variable indicating whether a product group in 1964 consisted of products which were relatively close substitutes in terms of capital and technical personnel requirements. The change in that measure between 1964 and 1970 (Δh) is assumed to pick up two types of movements over time. One is connected with the standardization of products during the product life cycle with respect to both embedded qualitative differences and production technology. We shall assume that the more intense the competitive pressure, the more rapid is the standardization and the more homogeneous will the commodity group become, i.e. Δh is lower the more rapid the standardization of the products within the group.

Another type of development affecting h is new products in the commodity group. Typically, "new" products are classified within certain miscellaneous commodity groups. Commodity groups which obtain many new products with technologies differing from the old products should be characterized by relative increases in the degree of heterogeneity. Countries with a comparative advantage in new products development ought consequently to be specialized in such commodity groups.

Suppose now that the purpose is to test if the product cycle theory can explain the observed negative specialization trends. Then, certain countries having had a comparative advantage in new product development should have lost (something of) that advantage. Other countries should have moved towards such an advantage from an advantage in products in the growth or mature phases of the life cycle. For both classes of countries it is expected that the change in specialization is oppositely related to P_{E64} and h_{64} compared to the specialization in 1964. Furthermore, the first group of countries should hypothetically increase their specialization on fast growing commodity groups and the second group on slow growing ones.

For certain countries, it was possible to investigate whether the intra-commodity group specialization in 1964 (P_k/P_E) and its change [$\Delta(P_k/P_E)$] had any systematic impact on the change in inter-commodity group specialization. Those two independent variables are included without specification of hypotheses regarding the sign of the regression coefficients. It was not even possible to motivate their inclusion from earlier developed trade theories. Since the inclusion of the two ton price ratio variables meant a loss of observations for several countries, regression results are in those cases only reported if those variables receive significance.

Table 9:7 presents the correlation coefficients for the main independent variables of the regression analysis of the 1964 and 1964-1970 changes in specialization. There is a tendency for commodity groups with a rapidly growing world market to have a high ton price in 1964 and a rapidly growing ton price 1964-1970. The latter two variables are positively correlated. Heterogeneous commodity groups in 1964 had smaller world markets and appear to have become relatively less heterogeneous in 1970. None of the correlation coefficients is so high as to pose a risk for multicollinearity in the following regression analysis. Before presenting that analysis it should be mentioned that regressions on another functional form have also been tried. Recall from chapter 6 that some regressions on the changing net export ratio also included the net export ratio of 1960. By analogy the 1964 net export ratio respectively the 1964 world export share were tested in regressions on the 1964-1970 changes in the respective two specialization measures. As might be expected from table 9:2 higher R^2 values were obtained in regressions including the 1964 levels. However,

Table 9:7

A correlation matrix for certain independent variables of the regression analysis

106 commodity groups

	Variable	1	2	3	4	5	6
1	$\Delta X_W/X_{W64}$	1.00[a]					
2	ΔP_E	0.19[a]	1.00[a]				
3	Δh	-0.02	0.08	1.00[a]			
4	P_{E64}	0.27[a]	0.44[a]	0.01	1.00[a]		
5	h_{64}	0.01	-0.08	-0.29[a]	0.09	1.00[a]	
6	X_{W64}	0.01	-0.05	0.06	-0.07	-0.30[a]	1.00[a]

[a] Significant at the 5% level (one-tail test).

since that inclusion only in exceptional cases altered the results for the other independent variables, we refrain from reporting these regressions in the following.

9.5.2 Trends in specialization for USA, Japan, West Germany and Great Britain

Table 9:8 presents the regressions for the USA and Japan on both measures of specialization and with respect to both the time period 1964-1970 and the year 1964. Those results can be compared to the corresponding results in chapter 8 for the year 1970.

In 1964, the US net export ratio was significantly, negatively related to the ton price. That relationship disappeared between 1964 and 1970, a change which may indicate a change in comparative advantage towards more technologically sophisticated products. That conclusion is unfortunately not supported by a significant positive sign of the regression coefficient for P_{E64} against $\Delta[(X-M)/(X+M)]$.

For the U.S. the change in specialization was more clearly related to the heterogeneity variable whose coefficient changed signs in regressions on $\Delta[(X-M)/(X+M)]$ and $[(X-M)/(X+M)]_{64}$. The product cycle theory appears to have received support as an explanation of the negative specialization trend. That support is strengthened by the fact that, according to both specialization measures, the U.S. specialization increased in commodity groups with rapidly growing world markets.

In 1964 as well as in 1970 the Japanese net export ratio was negatively related to the ton price while its world export share was positively related to the same ton price. Both measures were in 1964 larger the smaller total OECD exports were. This rather weak relationship disappeared between 1964 and 1970. Also with respect to h_{64} there are weak indications of opposite influences on the 1964 pattern of specialization and the trend in specialization in 1964-1970.

However, the Japanese case does not generate good support for either a factor proportions or product cycle account of the negative specialization trend. That is not surprising in view of the fact that there was a positive trend for Japan's world export share but rather astonishing from the point of view of the above reported finding of Caves & Uekusa [1976].

Table 9:8

Regressions on the changing net export ratios and world export shares 1964-1970 and on their 1964 levels. USA and Japan

106 commodity groups

Dependent variable	Constant	$\frac{\Delta X_W}{X_{W64}}$	ΔP_E	Δh	P_{E64}	h_{64}	X_{W64}	R^2	F-value (degrees of freedom)
	Regression coefficients (with standard errors) for								
USA:									
$\Delta(\frac{X-M}{X+M})$	-0.528	0.168^c (0.039)	-0.007 (0.012)	0.119 (0.185)	0.005 (0.006)	0.197^a (0.123)		0.204	5.132^c (5;100)
$(\frac{X-M}{X+M})_{64}$	0.634				-0.019^c (0.008)	-0.356^b (0.197)	0.005 (0.011)	0.095	3.573^c (3;102)
$\Delta(\frac{X}{X_W})$	-0.134	0.048^c (0.011)	-0.002 (0.003)	0.030 (0.052)	0.001 (0.002)	0.065^b (0.035)		0.199	4.978^c (5;100)
$(\frac{X}{X_W})_{64}$	0.182				0.001 (0.002)	0.007 (0.053)	0.007^c (0.003)	0.050	1.799 (3;102)
Japan:									
$\Delta(\frac{X-M}{X+M})$	0.186	0.031 (0.040)	-0.009 (0.012)	0.245^a (0.189)	0.001 (0.006)	-0.228^b (0.126)		0.070	1.495 (5;100)
$(\frac{X-M}{X+M})_{64}$	0.464				-0.12^a (0.008)	0.203 (0.201)	-0.015^a (0.011)	0.050	1.783 (3;102)
$\Delta(\frac{X}{X_W})$	0.046	-0.001 (0.008)	0.001 (0.003)	-0.013 (0.038)	0.000 (0.001)	-0.008 (0.026)		0.005	0.102 (5;100)
$(\frac{X}{X_W})_{64}$	0.056				0.005^c (0.002)	0.033 (0.035)	-0.003^a (0.002)	0.124	4.810^c (3;102)

Cont.

[a] Significant at the 10% level (one-tail test)

[b] " " " 5% " " " "

[c] " " " 2.5% " " " "

Legend:

$\Delta(\frac{X-M}{X+M})$ = change in 1964-1970 of the net export ratio of the country

$\Delta(\frac{X}{X_W})$ = change in 1964-1970 of the world market share of the country

$(\frac{X-M}{X+M})_{64}$ = net export ratio in 1964

$(X/X_W)_{64}$ = world market share in 1964

$\Delta X_W/X_{W64}$ = growth rate in total OECD exports 1964-1970

ΔP_E = change in ton price of exports of OECD-Europe 1964-1970

Δh = change in heterogeneity degree 1964-1970

P_{E64} = ton price of exports from OECD-Europe in 1964

h_{64} = the degree of heterogeneity in 1964

X_{W64} = OECD exports in 1964

According to their results Japan had in the post world war II period reversed its specialization pattern. Perhaps that finding could be interpreted to mean that the six year period studied here is too short in the Japanese case to reveal the possible long run influence of the trade determinants analysed in table 9:8.

Table 9:9 presents regression results for the two large European engineering exporters. Evidently, the two first regressions for the West German net export ratio were completely unsuccessful since no significant coefficients were obtained. In the following, we shall therefore abstain from commenting on those regressions.

The German world market shares were in 1964 higher for commodity groups with high ton prices (P_E) and low ton price ratios (P_{WG}/P_E). The change in its world market shares was oppositely related to both variables. Thus, there is evidence that these variables help explain the negative specialization trend for the German world export shares. West Germany seems to have moved towards a comparative advantage in technologically standardized commodity groups. It is interesting to note that the tendency of intra-commodity group specialization was in contrast to differentiate its exports on the high quality (= high ton price) assortment.

Even with respect to the growth rate of world exports, the German specialization moved in the opposite direction to that of the U.S. West Germany increased its specialization in slow growing commodity groups. That was also the case for groups with high ton price increases.

As was expected the British case offers good support for the hypothesis that the negative specialization trends are attributable to a change in comparative advantage. Not everybody would perhaps have assumed that it increased its specialization on technologically sophisticated commodity groups. That trend started from a very strong specialization in unsophisticated commodities in 1964 and was not sufficiently strong so as to make the 1970 pattern much different from the 1964 one. The trend is in any case consistent with the finding of chapter 3 that the relative salaries for British engineers were comparatively low.

At least according to the change in the net export ratio the British specialization increased during the period in slow growing products on the world market. World export shares increased more the more hetero-

Table 9:9

Regressions on the changing net export ratios and world export shares 1964-1970 and on their 1964 levels. West Germany and Great Britain.

106 commodity groups

Dependent variable	Constant	Regression coefficients (with standard errors) for $\frac{\Delta X_W}{X_{W64}}$	ΔP_E	Δh	$\Delta(\frac{P_k}{P_E})$	P_{E64}	h_{64}	$(\frac{P_k}{P_E})_{64}$	X_{W64}	R^2 F-value (degrees of freedom)
West Germany:										
$\Delta(\frac{X-M}{X+M})$	-0.148	0.018 (0.027)	0.006 (0.009)	0.004 (0.130)	0.010 (0.084)	-0.005 (0.004)	0.003 (0.092)	0.009 (0.076)		0.016 0.221 (7;98)
$(\frac{X-M}{X+M})_{64}$	0.664					-0.003 (0.004)	-0.108 (0.098)	0.017 (0.023)	-0.001 (0.005)	0.020 0.527 (4;101)
$\Delta(\frac{X}{X_W})$	-0.023	-0.009[a] (0.007)	0.005[c] (0.002)	0.039 (0.033)	0.014 (0.021)	-0.003[c] (0.001)	-0.017 (0.023)	0.032[b] (0.019)		0.185 3.181[c] (7;98)
$(\frac{X}{X_W})_{64}$	0.246					0.003[b] (0.001)	-0.013 (0.036)	-0.018[c] (0.008)	0.001 (0.002)	0.080 2.205[a] (4;101)

Cont.

Dependent variable	Constant	Regression coefficients (with standard errors) for $\frac{\Delta X_W}{X_{W64}}$	ΔP_E	Δh	$\Delta(\frac{P_k}{P_E})$	P_{E64}	h_{64}	$(\frac{P_k}{P_E})_{64}$	X_{W64}	R^2 F-value (degrees of freedom)
Great Britain:										
$\Delta(\frac{X-M}{X+M})$	-0.085	-0.070[c] (0.031)	-0.006 (0.010)	0.090 (0.148)		0.014[c] (0.005)	0.064 (0.099)			0.110 2.468[b] (5;100)
$(\frac{X-M}{X+M})_{64}$	0.618					-0.032[c] (0.005)	-0.024 (0.120)		0.001 (0.007)	0.290 13.856[c] (3;102)
$\Delta(\frac{X}{X_W})$	-0.064	-0.006 (0.006)	-0.002 (0.002)	0.108[c] (0.028)		0.003[c] (0.001)	0.039[c] (0.019)			0.214 5.448[c] (5;100)
$(\frac{X}{X_W})_{64}$	0.196					-0.004[c] (0.001)	-0.057[b] (0.033)		-0.001 (0.002)	0.105 3.982[c] (3;102)

[a] Significant at the 10% level (one-tail test)
[b] " " " 5% " (" " ")
[c] " " " 2.5% " (" " ")

Legend complementing those of table 9:8:

$(P_k/P_E)_{64}$ = ratio between the ton prices of country k and OECD-Europe in 1964

$\Delta(P_k/P_E)$ = change in 1964-1970 in the ton price ratio.

geneous the commodity groups were in 1964 or the more heterogeneous they became in 1964-1970. Hence, there are several indications that Great Britain moved towards a comparative advantage in new or young products and only one that it moved in the opposite direction. Again in the case Great Britain there are differences in results between the two measures of specialization.

In conclusion, we have demonstrated empirically at least for three of the four large countries, that the negative specialization trends might be attributable to changes in their comparative advantage. The results are far from being strong and unambigous. The most contradictory results are obtained for Japan which has elsewhere been shown to have reversed its pattern of specialization and on *a priori* grounds appears to be a likely case of an economy with systematically altered comparative advantage. Our findings are judged to be convincing enough for establishing a *temporary* conclusion that certain large industrial countries have experienced such thorough-going changes in their comparative advantage that their trade specialization patterns have tended towards a reversal. That temporary conclusion needs further testing for longer periods of time and for all tradable goods with an approach of the kind used in chapters 6 and 7 for Sweden.

Our results so far do not suffice to discriminate between factor proportions and product cycle accounts of the possible change in comparative advantage. However, the *changes* in comparative advantage which may underlie the results are not generally consistent with either Vernon's or Hirsch's suggestions as to which countries should have a comparative advantage on new or technologically young products.

9.5.3 Trends in specialization for Canada, France and Italy

In the year 1970 Canada, France and Italy had each about 7% of total OECD exports of engineering products. They can in this respect be classified as medium-large countries. Each of them increased net export ratios and world export shares. Their combined share of total OECD exports of engineering goods increased by about a third of the 1964 level.

For the three medium large and six small countries we shall report results on only one of the two specialization measures. Usually regressions on the net export ratio are chosen since they more often obtain the higher R^2 values.

Table 9:10 presents the regression results for Canada, France and Italy. Four of the six regressions are significant but none of the regressions has a high explanatory value.

For Canada there are rather weak indications of a changed comparative advantage. The regression coefficients for P_{E64} have different signs but are both insignificant. Stronger indications are obtained for the heterogeneity measures. Thus, from a certain specialization in heterogeneous commodity groups Canada increased its specialization in homogeneous ones or in those becoming relatively homogeneous. In doing this, Canada moved in the opposite direction of the USA. However, both countries increased their net export ratios more the higher the growth rate of world exports. In 1970 Canada was specialized in commodity groups with large world markets.

The regressions on the French patterns of specialization produced only one significant coefficient in each regression, namely that for P_{E64}. Since the sign of that coefficient was different in the two regressions that result could be interpreted as a weak support for the hypothesis that a changed comparative advantage accounted for the negative specialization trend of France. In fact, France reversed its 1964 pattern by 1970 from a highly significant but not very powerful specialization in technologically standardized commodity groups to a weakly significant one in technologically sophisticated groups in 1970.

The Italian case deviates from all other ones in that all results point to a strengthening of the 1964 specialization pattern in the following six year period. Thus, an initial specialization in standardized and homogeneous commodity groups was further increased. This result clearly opposes our hypothesis that the negative specialization trend for Italy is attributable to a change in Italy's comparative advantage. The intra-commodity group pattern of specialization lay similarly in relatively simple products. The Italien world export shares increased for slow growing commodity groups.

In conclusion, for two of the three medium large countries there are results which are consistent with our assumption that a change in the comparative advantage of a country accounts for its negative specialization trend. That the Canadian pattern moved in the opposite direction of the U.S. can be interpreted as a sign of

Table 9:10

Regressions on the changing net export ratio 1964-1970 and on its 1964 level. Canada, France and Italy

106 commodity groups

Dependent variable	Constant	Regression coefficients (with standard errors) for								R^2 F-value (degrees of freedom)
		$\frac{\Delta X_W}{X_{W64}}$	ΔP_E	Δh	$\Delta(\frac{P_k}{P_E})$	P_{E64}	h_{64}	$(\frac{P_k}{P_E})_{64}$	X_{W64}	
Canada										
$\Delta(\frac{X-M}{X+M})$	0.196	0.097[b] (0.054)	-0.003 (0.017)	-0.391[a] (0.257)		-0.006 (0.008)	-0.400[c] (0.172)			0.092 2.026[a] (5;100)
$(\frac{X-M}{X+M})_{64}$	-0.783					0.003 (0.007)	0.374[c] (0.166)		0.007 (0.009)	0.051 1.817[a] (3;102)
France										
$\Delta(\frac{X-M}{X+M})$	0.108	-0.041 (0.034)	-0.011 (0.011)	0.139 (0.156)	-0.059 (0.100)	0.010[b] (0.005)	0.041 (0.105)	0.050 (0.112)		0.066 0.994 (7;98)
$(\frac{X-M}{X+M})_{64}$	0.149					-0.013[c] (0.006)	0.001 (0.144)	0.020 (0.121)	-0.007 (0.008)	0.054 1.427 (4;101)
Italy										
$\Delta(\frac{X-M}{X+M})$	0.257	-0.038 (0.045)	0.004 (0.014)	-0.612[c] (0.213)		-0.001 (0.007)	-0.281[b] (0.142)			0.096 2.116[a] (5;100)
$(\frac{X-M}{X+M})_{64}$	0.437					-0.023[c] (0.007)	-0.349[c] (0.156)		0.013[a] (0.009)	0.157 6.306[c] (3;102)

[a] Significant at the 10% level (one-tail test)
[b] " " " 5% " " " "
[c] " " " 2.5% " " " "

Legend: See tables 9:8 and 9:9.

an increased division of labour between the two countries. Likewise, the French trend was similar to the British and opposite to both the German and the Italian ones.

9.5.4 Trends in specialization for six small industrial countries

Table 9:11 gives the regression results for six small industrial countries. Only four of twelve regressions are significant. The highest explanatory value is 0.242.

In 1964 Belgium was strongly specialized in technologically standardized, heterogeneous commodity groups but it increased its specialization in sophisticated commodities. Even for Belgium some support is received for the hypothesis that it is a change in comparative advantage that accounts for the negative specialization trend. The net export ratio grew more the slower the growth rate of world exports. In 1970 it was specialized in commodity groups with large world markets. The Belgian trend in specialization as regards technologically sophisticated products is similar to the British and French patterns and opposite the German and Italian ones.

Neither the 1964 nor the 1970 patterns of specialization was possible to explain for the Netherlands with the chosen regression approach. In contrast, it was according to table 9:11 easier to explain the change in specialization. Thus, the world export share (and net exports ratio) increased for commodity groups with high ton prices but decreased for those with the fastest growing ton prices. Since the latter changes were not very important and furthermore since ΔP_E was positively correlated with P_{E64}, the conclusion is that the Netherlands together with Belgium, France and Great Britain moved in the opposite direction to the German and Italian trends, namely towards an increased emphasis on sophisticated products. The results for the Netherlands differ, however, from those of the other countries in that we have not found indications that it experienced a change in comparative advantage that may account for its negative specialization trend.

The same conclusion holds also for the three smallest countries investigated, namely Norway, Denmark and Austria. Only one of the six regressions for these three countries is significant.

Table 9:11

Regressions on the changing net export ratios or world market shares and on their respective 1964 levels. Belgium, the Netherlands, Norway, Denmark, Austria and Switzerland

106 commodity groups

Dependent variable	Constant	Regression coefficients (with standard errors) for $\frac{\Delta X_W}{X_{W64}}$	ΔP_E	Δh	P_{E64}	h_{64}	X_{W64}	R^2	F-value (degrees of freedom)
Belgium-Luxemburg									
$\Delta(\frac{X-M}{X+M})$	0.032	-0.071^b (0.042)	0.015 (0.013)	0.044 (0.200)	0.015^c (0.006)	0.070 (0.133)		0.071	1.534 (5;100)
$(\frac{X-M}{X+M})_{64}$	-0.329				-0.026^c (0.007)	0.399^c (0.157)	0.007 (0.009)	0.171	7.028^c (3;102)
The Netherlands									
$\Delta(\frac{X}{X_W})$	0.003	-0.001 (0.003)	-0.004^c (0.001)	-0.001 (0.014)	0.001^c (0.000)	-0.007 (0.009)		0.181	4.413^c (5;100)
$(\frac{X}{X_W})_{64}$	0.034				0.000 (0.001)	0.008 (0.016)	0.001 (0.001)	0.011	0.374 (3;102)
Norway									
$\Delta(\frac{X-M}{X+M})$	0.082	-0.015 (0.032)	0.015^a (0.010)	-0.131 (0.151)	-0.003 (0.005)	-0.105 (0.101)		0.040	0.825 (5;100)
$(\frac{X-M}{X+M})_{64}$	-0.482				-0.009^a (0.006)	-0.135 (0.134)	-0.010^a (0.008)	0.044	1.558 (3;102)

Cont.

Dependent variable	Constant	$\frac{\Delta X_W}{X_{W64}}$	ΔP_E	Δh	P_{E64}	h_{64}	X_{W64}	R^2	F-value (degrees of freedom)
	Regression coefficients (with standard errors) for								
Denmark									
$\Delta(\frac{X-M}{X+M})$	0.061	0.033 (0.046)	0.014 (0.014)	-0.240 (0.218)	-0.007 (0.007)	-0.260^b (0.146)		0.057	1.209 (5;100)
$(\frac{X-M}{X+M})$	0.023				-0.010^a (0.008)	-0.0861 (0.189)	-0.006 (0.011)	0.020	0.704 (3;102)
Austria									
$\Delta(\frac{X-M}{X+M})$	0.012	0.016 (0.049)	-0.001 (0.015)	0.398^b (0.233)	-0.003 (0.007)	0.023 (0.155)		0.031	0.642 (5;100)
$(\frac{X-M}{X+M})64$	-0.297				-0.009 (0.008)	0.333^b (0.186)	-0.011 (0.010)	0.061	2.224^a (3;102)
Switzerland									
$\Delta(\frac{X}{X_W})$	0.017	-0.011^c (0.002)	0.000 (0.001)	-0.039^c (0.011)	-0.000 (0.000)	-0.008 (0.008)		0.242	6.399^c (5;100)
$(\frac{X}{X_W})64$	0.020				0.001^a (0.001)	0.011 (0.014)	0.000 (0.001)	0.026	0.905 (3;102)

[a] Significant at the 10% level (one-tail-test)
[b] " " " 5% " " " "
[c] " " " 2.5% " " " "

Legend: See tables 9:8 and 9:9.

The Swiss trend in specialization is explained to about 24%. Unfortunately, there are at best weak indications supporting the hypothesis that the negative specialization trend is attributable to a change in comparative advantage. In 1964 Switzerland had a higher world export share the more technologically sophisticated the commodities. That relatively weak relationship disappeared by 1970. In addition, the changing world export share was negatively related to Δh and also negatively, but insignificantly to h_{64} while $(X_{SZ}/X_W)_{64}$ was positively, but insignificantly related to h_{64}. Instead, the Swiss specialization trend was most clearly (and negatively) correlated with the growth rate of the world market. Perhaps even more than Canada, Switzerland had the opposite specialization trend as the U.S. The correlation coefficient between $\Delta(X_k/X_W)$ of the two countries was -0.31, which is significant at the 5% level.

In conclusion, there are fewer indications of a changing comparative advantage for small and medium small countries than for medium large and large ones. Sweden is in this context a medium small country being as large as Belgium in terms of its world export share of engineering goods, surpassing the other five industrial countries of table 9:11. Next we shall investigate the Swedish trends and compare the results with both the inter-industry trends of chapter 6 and the trends for other industrial countries.

9.5.5 The Swedish trends in specialization

The regression analysis for Sweden includes a third measure of specialization, the home market share, H/C. That measure was used in the industry analysis of chapters 4 and 6 which is a reason for its inclusion also in the commodity group analysis. That measure utilizes certain other independent variables to be discussed separately below.

Table 9:12 presents the results with respect to the net export ratio and the world export share. Recall from chapter 6 that Sweden moved from a capital intensive towards a human capital intensive pattern of industry specialization. That would be consistent with a commodity group specialization developing from a low ton price towards a high ton price pattern The net export ratio of 1964 is in fact negatively related to P_{E64}

Table 9:12

Regressions on changes in Swedish net export ratios and world export shares 1964-1970 and on their 1964 levels
106 commodity groups

Dependent variable	Constant	Regression coefficients (with standard errors) for								R^2 F-value (degrees of freedom)
		$\frac{\Delta X_W}{X_{W64}}$	ΔP_E	$\Delta(\frac{P_S}{P_E})$	Δh	P_{E64}	$(\frac{P_S}{P_E})64$	h_{64}	X_{W64}	
$\Delta(\frac{X-M}{X+M})$	0.0121	-0.0491 (0.0459)	0.0087 (0.0142)	0.0017 (0.0316)	0.1073 (0.2255)	0.0028 (0.0068)	0.0541 (0.0740)	-0.0777 (0.1817)		0.026 0.037 (7;98)
$(\frac{X-M}{X+M})64$	0.1251					-0.0226^c (0.0077)	-0.0764 (0.0825)	0.1041 (0.2054)	-0.0075 (0.0103)	0.085 2.352^a (4;101)
$\Delta(\frac{X}{X_W})$	0.0025	-0.0043^b (0.0022)	-0.0009 (0.0007)	0.0005 (0.0015)	0.0163^a (0.0106)	0.0001 (0.0003)	0.0021 (0.0035)	-0.0010 (0.0086)		0.099 1.532 (7;98)
$(\frac{X}{X_W})64$	0.0312					-0.0004 (0.0005)	0.0053 (0.0055)	-0.0047 (0.0138)	0.0000 (0.0007)	0.017 0.438 (4;101)

[a] Significant at the 10% level (one-tail-test)
[b] " " " 5% " " " "
[c] " " " 2.5% " " " "

Legend: See tables 9:8 and 9:9.

but the positive sign of the relationship between $\Delta[(X-M)/(X+M)]$ and P_{E64} is not significant. The corresponding regression coefficients in the regressions on $(X/X_W)_{64}$ and $\Delta(X/X_W)$ have also the right sign but are both insignificant. Thus, there are only at best very weak indications of a specialization trend at the commodity group level which is similar to that at the industry level. Furthermore, it is not possible to say whether this difference in results between chapter 6 and the present chapter is attributable to the shorter time period studied or to the different methodologies used.

Also with respect to some of the other variables $(P_S/P_E)_{64}$ and h_{64}, different, but insignificant signs are received in the first two regressions of table 9:12. The results of that table apparently do not give much support to the assumption that the specialization trends for Sweden are attributable to changes in its comparative advantage. On this point the Swedish results are comparable with those of several other small countries. Despite this fact the industry analysis of the longer period 1960-1970 did generate substantial support for that assumption. It appears as if the Swedish trade adjustment was not one of the most powerful or required a longer time period than that of other industrial countries.

Chapter 6 indicated that the Swedish engineering industry increased its specialization in slow growing domestic markets. According to table 9:12 the Swedish world market shares increased in commodity groups with low world market growth rates. In fact, an inclusion of $[(X-M)/(X+M)]_{64}$ in the first regression makes the negative coefficient for $\Delta X_W/X_{W64}$ significant. Therefore, we may conclude that the commodity group and the industry analyses lead to the same conclusion on this point, namely that Sweden increased its specialization in slow growing markets.

In contrast to the two other specialization measures, the home market share could be obtained for the whole period 1960-1970. Section 9.3 demonstrated that a negative specialization trend appeared at the commodity group level in contrast to the industry level. The following regression analysis aims at seeking explanations to that negative trend. However, only two of the independent variables cover the whole period, namely

$\Delta C/C_{60}$ and C_{60}. The other independent variables were only obtainable for the period 1964-1970 and 1964 respectively. Therefore, we have to assume that the latter variables are representative to the full period 1960-1970 and to the levels in 1960, respectively.

Some regression results are presented in table 9:13. Also at the commodity group level the 1960 home market share is possible to explain to a large extent with the regression approach (R^2=0.313)[4]. The home market share was higher for commodity groups which (in 1964) were capital intensive/technical personnel extensive, heterogeneous and had a low tariff level. Those results are strikingly similar to those found at the industry level.

The trend in H/C is not as well explained. But there are certain results which might help to explain the negative specialization trend for H/C. Different signs are obtained in the two regressions for P_{E64}, $(P_S/P_E)_{64}$ and h_{64} although significance is received only in one of the two regressions. Furthermore, $\Delta(H/C)$ is negatively related to Δh as opposed to the positive relationship between $(H/C)_{60}$ and h_{64}. Including also $(H/C)_{60}$ as an independent variable in the first regression led to a significant negative coefficient for $\Delta C/C_{60}$.

In conclusion, although the negative specialization trend was weaker in terms of H/C for the longer period 1960-1970, than of both $(X-M)/(X+M)$ and X/X_W for the period 1964-1970, the former trend was better explained by the approach followed here. That result together with the industry results of chapters 6 and 7 suggest that the trade adjustment processes connected with a possible change in Sweden's comparative advantage are long run processes hardly noticable over a period almost as short as a business cycle. Alternatively, the change must be more thorough-going or the economies more adaptable in other industrial countries, for which the signs of changes in comparative advantage were substantially more significant in the period 1964-1970.

9.6 Conclusions

Chapter 9 sought to investigate whether or not results for Sweden in chapter 6 could be generalized to lower

Table 9:13

Regressions on the change in Swedish home market shares 1964-1970 and on its 1960 level

106 commodity groups

Dependent variable	Constant	Regression coefficients (with standard errors) for										R^2 F-value (degrees of freedom)
		$\frac{\Delta C}{C_{60}}$	ΔP_E	$\Delta(P_S/P_E)$	Δh	Δt	P_{E64}	$(P_S/P_E)_{64}$	h_{64}	C_{60}	t_{64}	
$\Delta(\frac{H}{C})$	0.1417	-0.0018 (0.0415)	0.0044 (0.0078)	0.0180 (0.0174)	-0.2274[b] (0.1243)	-1.2500 (1.0507)	0.0034 (0.0036)	0.0618[a] (0.0415)	-0.0761 (0.1018)			0.073 0.946 (8;96)
$(\frac{H}{C})$	0.7560						-0.0240[c] (0.0039)	-0.0142 (0.0423)	0.1798[b] (0.1033)	0.0000 (0.0000)	-1.3185[a] (1.0094)	0.313 9.031[c] (5;99)

[a] Significant at the 10% level (one-tail-test

[b] " " " 5% " " " "

[c] " " " 2.5% " " " "

Legend (complementing those of tables 9:8 and 9:9):

C/C_{60} = growth rate of domestic consumption 1960-1970

Δt = the tariff decrease (on total Swedish imports) 1964-1970

C_{60} = domestic consumption in 1960

t_{64} = Swedish import tariff (on total imports in 1964.

levels of aggregation and to other industrial countries. First and foremost, the aim was to detect negative specialization trends and to seek explanations for them.

The overwhelming impression of this chapter is that all industrial countries seem to have negative specialization trends both at the commodity group level and also within the commodity groups at least for the nine countries possible to study. Of course, no country was found to have reversed its trade specialization between 1964 and 1970. Therefore, the question remains whether the negative trends should be interpreted as indications of parts of reversals of trade patterns or as parts of tendencies towards an evening out of the trade patterns and the underlying differences in comparative advantage. The latter result would within the factor proportions theory be consistent with a tendency towards equalization of (relative) factor prices or of relative factor endowments.

In the Swedish case the negative specialization trends were found to be more significant at the commodity group level than at the industry level. On the other hand, the indications of a change in Sweden's comparative advantage towards human capital intensive production were weaker at the former level possibly in part due to the shorter time period covered. The Swedish readjustment process appears in fact not to be one of the more powerful ones among the 14 industrial countries. In addition, Sweden did not move at all strongly towards more technologically sophisticated products compared to the other countries moving in the same direction.

Methodologically, chapter 9 established a new direction of study in its explicit inclusion of determinants of the product cycle theory. However, the latter theory was tested on the change in trade patterns rather than the usual procedure of testing it on the patterns of trade in a given year. Since no hypotheses could be formulated regarding which countries should move upwards the product life cycle in their specialization and which ones would move downwards, the hypotheses testing was simplified to investigate for which countries that theory might offer explanations of the negative specialization trend. The conclusion was that the product cycle theory and/or the factor proportions theory could explain that trend for certain large and medium large countries but not for several small ones. In fact, the two theories were only in some respects distinguishable with the method used. In any case, the method to distinguish operationally between separate elements of the product cycle theory in order to facilitate its testing on changes in specialization in a cross-section of products or for individual products.

The difference in results between large and small countries may be taken as an indication that the driving forces behind the specialization trends are to be sought in the comparative growth patterns of the large countries. However, the comparative static framework of the empirical analysis do not allow any conclusions on the direction of causality. Rather the results should be taken as a hypothesis put forward for further testing.

Footnotes

1. 105 commodity groups for the home market share for reasons discussed in chapter 8.
2. Except for the Canadian world export shares, for which no relationship at all was found.
3. To some extent the high R^2 value for Italy is explained by one extreme observation. Even after exclusion of that observation a negative sign for $(P_I/P_E)_{64}$ is received as demonstrated by the following regression:

$$\Delta(P_I/P_E)_{64-70} = 0.803 + \underset{(0.092)}{0.062}\,h_{64} - \underset{(0.004)}{0.015^b}P_{E64} - \underset{(0.067)}{0.794^b}(P_I/P_E)_{64} -$$

$$- \underset{(0.380)}{0.564^a}(X_I/X_W)_{64}; \qquad R^2 = 0.608 \quad F(4;99) = 38.304$$

[a] Significant at the 10% level (one-tail test).
[b] " " " 1% " " " "

According to this regression the ton price ratio increased more for commodity groups in which Italy had a low world market share in 1964. That variable was for other countries not significant.

4. The inclusion of the trade exposure measure, r, discussed in chapter 8, was also tried for 1960. The same results were obtained. It was highly influential and reduced the negative coefficient for P_{E64}.

Chapter 10

SUMMARY AND CONCLUSIONS

The present study has had three main purposes. One has been to present empirical results regarding the engineering trade specialization of Sweden and other industrial countries. A second purpose has been to contribute to further theoretical and empirical investigation of the determinants of foreign trade. Thirdly, the study has presented new methodologies for the study of trade specialization patterns.

This concluding chapter summarizes and interprets the results of the study in accordance with these purposes. The first section discusses the empirical findings for Sweden and provides some additional tests aimed at generalizing earlier results. The second section discusses the findings for other industrial countries. Next follows a section which compares factor proportions theories with the product cycle theory and which briefly summarizes the new methodologies of the study. A concluding section discusses directions for further study.

10.1 The Swedish engineering trade patterns

Probably the most important finding from a policy point of view was that Sweden increased its specialization in the 1960's in engineering products in which it was not specialized at the beginning of the period. Studies using different specialization measures and different levels of aggregation rule out the possibility that the negative specialization trend could be a *pure* intra-industry trade phenomenon. Instead it appears as if the forces behind the changes in inter-industry specialization also acted on the intra-industry specialization pattern. Table 10:1 provides a summary of all results showing a negative specialization trend.

Chapter 6 of the study investigated whether there was a factor proportions explanation of the observed trend for the industry net export ratio. Tariffs alone could not explain that trend and neither could technical change.

During the period two of the factor intensities stud-

Table 10:1

Results of the study showing negative specialization trends in Swedish engineering trade

Aggregation level	Period	Specialization measure
34 industries	1960-1970	$\frac{X-M}{X+M}$ (but not for $\frac{H}{C}$)
106 commodity groups	1964-1970	$\frac{X-M}{X+M}$, $\frac{X}{X_W}$
106 commodity groups	1960-1970	$\frac{H}{C}$
Within the 106 commodity groups	1964-1970	Other methodology used (cf section 10.4)

Legend: $(X-M)/(X+M)$ = the Swedish net export ratio (X = exports, M = imports)

X/X_W = the Swedish share of total OECD exports (X_W = OECD exports)

H/C = the Swedish home market share (H = gross output minus exports, C = domestic consumption or $H+M$)

ied increased substantially, namely the capital and technical personnel intensities, and one decreased somewhat, namely the skilled manual worker intensity. For the two former intensities it was possible to analyse whether *technical* improvement was also associated with a *technological* bias which caused changes in the relative differences of the factor intensities of the industries. These tests gave strong support for a factor proportions approach based on an assumption of a long run stability in the relative differences of the physical factor intensities. These differences remained very much the same throughout a 15-year long period, the exceptions being relatively small changes for industries with similar values for the two factor intensities.

Ruling out the possibility that the tariff change and the technological change alone could explain the negative specialization trend, there remained two possible explanations within a factor proportions framework.

One was that the relative commodity prices of the world market had changed systematically in favour of goods in which Sweden did not have a comparative advantage in 1960. As a small, price-taking economy it would have had in that case to adjust by changing the trade specialization and factor supplies to comply better with the new relative prices and in consequence move away from the earlier pattern of specialization.

Some indications supported the assumption that relative prices developed in favour of technical personnel or skill intensive goods and against capital intensive and skill extensive goods, an explanation which could not be tested explicitly. Therefore, it cannot really be judged whether or not the specialization change was externally determined.

A fourth possible explanation was a Rybczynski-type change in Sweden's factor abundance and comparative advantage. The Rybczynski theorem is derived from a model with only two goods, two factors and two countries. It rests on a small country assumption implying constancy of the relative commodity price. Our approach was a multicommodity-multifactor-small country one. However, we abstained from assuming homothetic demand functions, i.e. we allowed for different income elasticities for the products although they were assumed to be the same for Sweden and the rest of the world.

Within this framework it was shown that a change in Sweden's comparative advantage might account for the negative specialization trend. From a (weak) capital intensive specialization in 1960, Sweden specialized less in capital intensive and non-skilled labour intensive production and more in technical personnel and skilled manual worker intensive production. Hence, in 1970 the Swedish pattern was (weakly) positively related to the latter intensity.

The Swedish industry specialization increased during the period in production which had a slow domestic market growth and small tariff decreases. The former result also holds for specialization measures other than the net export ratio and at the commodity group level of aggregation. Unfortunately, the results for the factor intensities were not as easy to generalize at that level. First of all a proxy variable had to be substituted for the factor intensities. That variable, the metric ton price, was only weakly (positively) related to the skilled manual worker intensity.

But its strong negative correlation with the capital intensity and equally strong positive correlation with the technical personnel intensity made it a satisfactory measure for those two intensities. Thus a decreased comparative advantage in capital intensive and an increased one in technical personnel intensive production should be possible to study.

A second limitation of the commodity group analysis was the shorter time period, 1964-1970 instead of 1960-1970. Nevertheless, that analysis made possible a comparison between the specialization patterns of Sweden and those of other industrial countries. The following are the main findings.

First, Sweden was specialized in technologically standardized, i.e. capital intensive/technical personnel extensive commodity groups in 1964[1]. The change in specialization was positively but not significantly related to the metric ton price. Hence, it is not possible to conclude that Sweden's change in *commodity group* specialization 1964-1970 was characterized by a development towards a more technologically sophisticated pattern. The only clear indication for the period 1964-1970 of a decreased comparative advantage in capital intensive low-skilled labour intensive production was obtained for the intra-commodity group specialization. The latter developed more towards a high quality mix within product groups, with, according to the industry analysis, a weakened comparative advantage.

Comparisons with other industrial countries were undertaken to reveal in which respects the Swedish patterns deviated from those of other countries. The Swedish engineering industry had together with Switzerland and West Germany a relatively large proportion of commodity groups within which it was specialized in a high quality mix. On the other hand, it was one of the countries with the most pronounced specialization in technologically standardized commodity groups. Moreover, it did not have a significant trend away from such a pattern in the period 1964-1970.

Both the industry and commodity group trends indicated that factor proportions accounts worked better for relatively long periods of time. As some information suggests that there was substantial excess demand for engineers and skilled manual workers (and perhaps also financial capital) in a period when relative wages remained unchanged, the incentive structure may have been unclear for domestic firms. Different groups of engineering firms may have specialized differently. For instance, the existing shortage of engineers and skilled manual workers at going relative wages may have prevented new firms from specializing in production intensive in such labour skills and at the same time may have established an exit barrier for existing firms. Other factor or commodity market barriers may have similarly delayed or distorted the production adjustment of the Swedish economy.

For these reasons one chapter was designed to compare the patterns of specialization changes for new firm entries, diversification entries, permanent firms[2], independent exits and specialization exits. These changes were defined so as to exhaust the change in Sweden's domestic consumption save for a residual factor, namely the change in the net exports ratio of domestic consumption, $\Delta[(X+M)/C]$.

The results showed that the diminished Swedish trade specialization in fast growing industries (and commodity groups) could be attributed entirely to the development of permanent firms. In contrast, entries of both categories of new firms specialized in fast growing markets and both types of exits were relatively smaller in such markets. It appears as if existing Swedish engineering firms stuck to their existing product mix which happened to be in markets with a relatively low growth rate. Their efforts to change product mixes by diversification entries and specialization exits were not substantial enough to balance the development of permanent plants, expansion entries and contraction exits.

The results for especially the technical personnel intensity but also the skilled manual worker intensity suggest that the factor markets functioned imperfectly in the 1960's. Only permanent firms increased their specialization in technical personnel and skilled manual worker intensive production. Both new firm and diversification entries had in contrast an unfavourable growth in technical personnel intensive industries. Moreover, their comparative growth performance was not correlated with the skilled manual worker intensity. Combined with the registered excess demand for both factors, those results suggest the existence of determinants which tended to distort or delay the Rybczynski-type adjustment mechanisms of a hypothetical change in Sweden's comparative advantage.

The necessary use of domestic consumption instead of gross trade in the specialization measures of the five categories of firms cast doubt on the results with respect to the capital intensity. Various results indicated that transport costs were especially large for the most capital intensive engineering products and in effect sufficiently large to establish effective barriers to both exports and imports.

The analysis of entry and exit barriers also showed that a high initial concentration ratio of domestic producers combined with a low import share of domestic consumption prevented or delayed substantially both the entry of new domestic firms and foreign firms. Although the inclusion or exclusion of the adjusted four firm concentration ratio did not appear to affect the signs of the relationship between the trade specialization change and the factor intensities, the mere existence of that entry barrier for foreign producers probably diminished the explanatory power for those intensities. In other words, that barrier should have tended to prolong the adjustment period of domestic permanent producers.

Summarizing, several results suggest that a decade long period may only be long enough to reveal the trend in adjustment following a possible change in comparative advantage. Given that there is no alteration in adjustment incentives a completion of the adjustment should require a substantially longer time period. This conclusion assumes implicitly that the same incentives exist for the whole tradable goods sector in Sweden and not only for the engineering industry, a proposition that is investigated in a forthcoming study of the trends in specialization of Swedish regions[3]. Suffice it to include here certain results indicating whether the negative specialization trend (in the net export ratio) holds more generally. In the latter study 112 industries are classified according to degree of trade exposure in 1960 and 1970. Fifteen were found to have very little exports and imports compared to production and consumption, respectively. Of the remaining industries, seventeen were basically raw material intensive ones. Hence, 80 industries accounting for 66% of manufacturing employment in 1970 were classified as international *and* footloose industries. The following cross-sectional relationship between the change in the net export ratio 1960-1970 (y_{60-70}) and the 1960 net export ratio (x_{60}) was estimated for those industries:

$$y_{60-70} = -0.41 - 0.296x_{60} \qquad R^2 = 0.275$$
$$F(1;78) = 29.611$$

to be compared with the corresponding relationship for the 34 engineering industries:

$$y_{60-70} = 3.70 - \underset{(0.103)}{0.380}x_{60} \qquad R^2 = 0.297$$
$$F(1;32) = 13.513$$

Two conclusions follow. First, there is a negative specialization trend for the whole non-raw material based, tradable goods manufacturing industry. In fact, it holds also for all 112 industries. Secondly, the negative trend is stronger for the engineering sector than for all 80 industries in terms of the magnitude of the slope coefficient.

The conclusion that the trade adjustment process is time-consuming also hangs on the stated assumption that the same incentives for specialization changes prevail for a longer period than the two periods studied here, viz. 1960-1970 and 1964-1970. At the industry level an extension of the period is now possible to the years 1960-1974. The following regression result was received for that period and for the 80 tradable goods industries:

$$y_{60-74} = 2.38 - \underset{(0.061)}{0.395}x_{60} \qquad R^2 = 0.348$$
$$F(1;78) = 41.642$$

Obviously, the extension of the period increases the explanatory power and the magnitude of the slope coefficient. Both results support a conclusion that the same underlying forces behind the trade adjustment trend of the 1960's were at work also during the first half of the 1970's.

A further extension of the period back to the 1950's required a change from the industry classification system to a commodity group (SITC) one. Utilizing a key between SITC-Revised and the earlier SITC system it was possible to obtain data for the engineering sector at the 3-digit level of SITC-Revised and the period 1953-1974. Excluding only aircraft, data for 27 commodity groups were put together for the years 1953, 1960, 1967 and 1974. Three time periods have been studied, viz. a 7-year period 1953-1960, a 14-year period 1953-1967 and a 21-year period 1953-1974. The specialization trends for the three periods are shown in figure 10:1.

Obviously, there is a negative specialization trend in each period. Three points should be emphasized. First, the negative slope coefficients fall from -0.60

Figure 10:1

Specialization trends in 1953-1974 for three periods and for 27 commodity items in the engineering sector

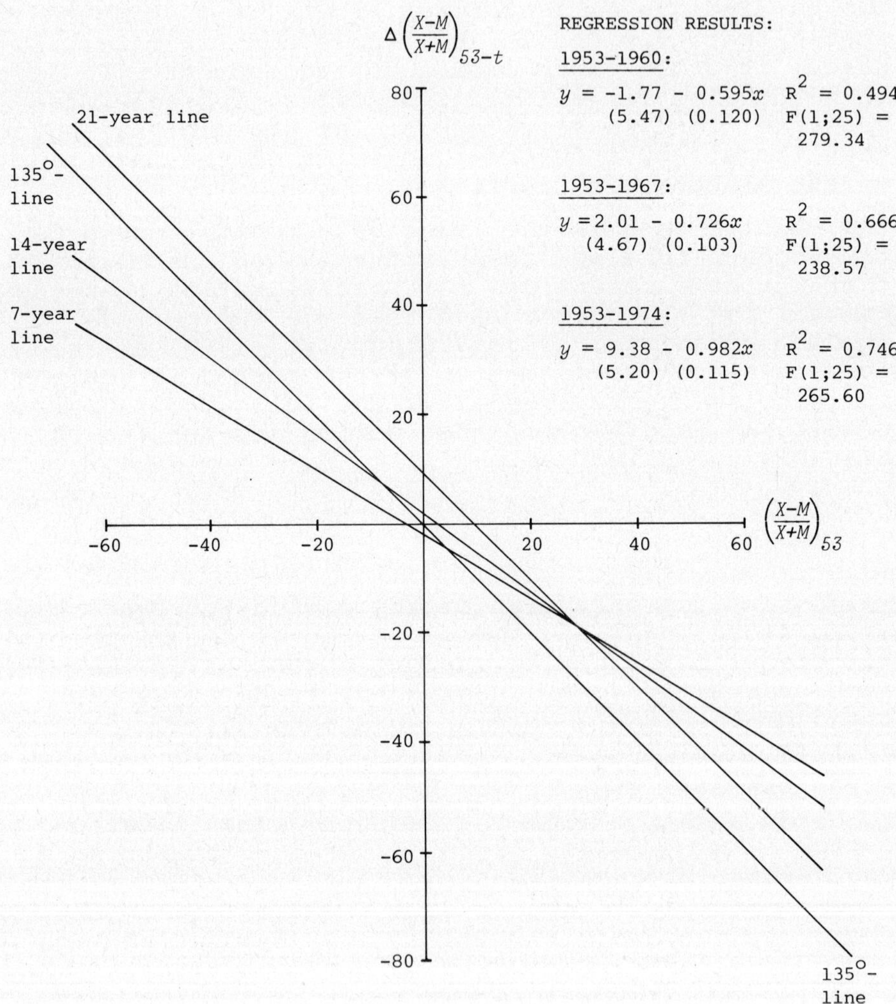

Note: The 27 commodity items are the 3-digit SITC-Revised items within SITC 7 (excl 734) plus 812 and 861.

to -0.73 and further to -0.98, i.e. the specialization trend is a very prolonged one and appears to be the same throughout the twentyone years. Secondly, not even the longest of the periods generates a lower slope coefficient than the 135°-line, despite the fact that 11 of the 27 observations actually experienced a change in sign of their net export ratios. It is, however, so close that another extension of the period with seven years would appear to be sufficient to create a slope coefficient significantly below -1.0.

Thirdly, the multiple correlation coefficients increase successively. The change in the net export ratio between 1953 and 1974 is explained up to 75% by 1953 net export ratio.

Although, figure 10:1 does not show a reversal of the 1953 specialization pattern it is a sufficiently strong demonstration to support the assumption that the negative specialization trends are not primarily indications of increased two-way trade. Instead, our earlier (temporary) conclusion that Sweden has in fact experienced a systematic and thorough-going change in its factor abundance and comparative advantage gains further strength. Of course, such a change should also give rise to intra-industry trade. Hence, the conclusion would be that we might have found an important link between inter- and intra-industry trade specialization patterns. If the determinants behind the changing inter-industry specialization persist long enough we would expect a shift in Sweden's trade specialization compared with the pattern of the 1950's.

10.2 Engineering trade patterns of other industrial countries

Since it was impossible to obtain the same industry classification system for all 13 other industrial countries, the analysis was undertaken at the commodity group level for the period 1964-1970. The short period, the methodology available and the fact that it was not possible to define each country's factor abundance restricted the ambitions of that analysis compared to that for Sweden. Apart from throwing some light on how Sweden's international specialization patterns related to the corresponding patterns of other industrial countries, the comparisons had three aims. One aim was to investigate whether the negative specialization trends held only for Sweden or more generally. An additional aim was to study whether trends for other countries could be attributed to changes in comparative advantage of the kind observed for Sweden. A third aim was to demonstrate the possible usefulness of a new methodology for inter- and intra-commodity group specialization analysis. That methodology is further discussed in the next section.

All 13 industrial countries had negative specialization trends, in most cases for both measures of specialization studied. Sweden was far from having the strongest trend. Morever, for nine of the countries it was possible to study the corresponding trend *within* the commodity groups. It turned out that each one of those countries had a negative intra-commodity group specialization trend. Hence, it seems safe to conclude that practically all industrial countries experienced profound specialization changes similar to Sweden.

The analysis of whether these trends were associated with changes in the factor abundance of each country was limited, as only two kinds of changes could be captured, because of the properties of the technology proxy used, (i.e. the metric ton price). The factor abundance must have altered from a physical capital abundance/technical personnel scarcity situation towards a physical capital scarcity/technical personnel abundance situation. Naturally the six year period could not be expected to contain the whole change. Therefore, the countries can be divided into three groups as regards the obtained results on this point. One group consists of countries with different signs on the regression coefficient for the technology proxy in regressions on the change in specialization and the 1964 level of specialization. West Germany, Great Britain, France and Belgium belonged to that group. The latter three countries were in 1964 specialized in capital intensive/technical personnel extensive commodity groups but decreased their specialization in such groups 1964-1970. All three countries also had an intra-commodity group specialization in low quality (or capital intensive/technical personnel extensive products). West Germany underwent the opposite change and had in addition the opposite type of intra-commodity group specialization.

A second group of countries are those for which the two signs of the regression coefficient for the technology proxy were different but only one of them significant. That group included USA, Japan and Switzerland. Sweden can also be included in this group. The two former countries (and Sweden) had larger net export ratios in 1964 the more capital intensive/technical personnel extensive the commodity groups were, while Sw had the opposite specialization. In 1964-1970 they at least did not strengthen their 1964 specialization.

The remaining group of countries is a miscellaneous group. It includes two countries with both coefficients non-significant (Canada and Austria), three countries with negative coefficients (Italy, Norway and Denmark) and one country with positive coefficients (the Netherlands).

In conclusion, there are some results that may indicate a change in comparative advantage especially for the larger countries. Moreover, in Europe it appears as if West Germany altered its specialization in the opposite direction to Great Britain, France, Belgium and the Netherlands. Therefore, the results although not at all powerful suffice to establish a hypothesis of major changes in the factor abundance between large and medium large countries and also Belgium and Sweden. The negative specialization trends of the other small countries may be repercussious from trade changes of the large countries without connection to alterations of their own factor endowments or prices.

The technology proxy used (the metric ton price) is a variable which works equally well within a product cycle framwork. In fact, the methodology of the commodity group analysis contained other variables, which were operationalizations of trade determinants specific to the product cycle theory (cf section 10.3 below). As was the case for the factor proportions theory, the tests of the product cycle theory could not originate from specified hypotheses regarding each country's comparative advantage and its possible change. Instead, the ambition was merely to investigate whether the product cycle theory might explain the negative specialization trends for some countries.

The US and Great Britain were in 1964 specialized in relatively homogeneous commodity groups with capital intensive/technical personnel extensive technologies. They both diminished their specialization in such commodity groups. Canada specialized in 1964 in heterogeneous commodity groups but increased its specialization in homogeneous ones after that year. All three countries are cases in which the product cycle theory might help explaining the negative specialization trend.

For several other countries the heterogeneity variable proved to be influential for the patterns of specialization but without clear evidence on or even evidence against a product cycle account of the negative specialization trends. Both that account and the factor proportions theory assign the market growth rate a potentially powerful rôle as a determinant of a country's specialization change. Thus, Sweden was shown to decrease its specialization in fast growing markets (measured by the growth rates of domestic consumption or world exports). West Germany, Great Britain, Belgium and Switzerland had the same tendency to diminish their specialization on commodity groups with rapidly growing world exports. The U.S. moved strongly in the opposite direction and so did Canada. Obviously, the results do not appear to be predictable from only the size and development level of the countries as originally proposed by Hirsch [1967] or Vernon [1966].

Summarizing, the most important finding of the analysis of engineering trade specialization of 13 industrial countries is that practically all of them had negative specialization trends. There are also some indications of changes in comparative advantage which are systematically related to the technology and/or the heterogeneity of the commodity groups. Hence, there is enough evidence from the study to put forward the hypothesis that several of the countries have experienced a general change in comparative advantage which might be attributable to changing relative factor endowments or prices.

10.3 Theoretical and methodological conclusions

Compared to other empirical studies of the factor proportions theory this study deviates in several respects. First of all, it is strictly based on multiproduct-multifactor-two country models in which the country in question is treated as a small country facing exogenous world market prices and technologies. The demand side and certain trade impediments are explicitly taken care of both in the construction of dependent variables and the choice of independent variables. The analysis includes investigation of trade patterns in a given year as well as changes in trade patterns in periods of 6 and 10 years. Despite the fact that the dependent variables are bounded, practically all relationships are found to be linear[4].

Almost all results show that the pattern of specialization in a given year is only weakly associated with the factor intensities. That is a rather common result in earlier published factor proportions studies. Our study includes special efforts to find out whether the low explanatory power can be attributed to the particular definitions of the independent variables or to aggregation problems. They were not! Instead, the results proved to be more sensitive to the choice of the specialization variable. Specifically, the factor intensities were powerful explanatory variables in regressions on the home market share but not on the net export ratio. Unfortunately, this result is probably attributable to varying tradability and its close relationship with factor intensities.

Substantially more powerful results were obtained in regressions on the change in specialization. It appears, in consequence, that a multiproduct-multifactor analogue of the Rybczynski theorem receives more support than the corresponding analogue of the Heckscher-Ohlin theorem. If in fact the factor abundance of a given country undergoes a major change the latter result is not surprising for years lying within the adjustment period. For Sweden it has been shown that the structural adjustment in a nine-year period includes elements of a perhaps medium-run nature which elements delay or distort the trade and production adjustment proposed by the Rybczynski theorem. Important entry and exit barriers were revealed.

By and large, the kind of factor proportions models underlying our analysis of specialization changes do seem to be fairly realistic in the Swedish case save for certain unrealistic assumptions of perfectly functioning domestic factor and commodity markets. This may appear astonishing since the Rybczynski (as well as the Hecksher-Ohlin) theorem has not yet lent itself to generalization beyond the 2x2x2 models. However, if the underlying incentives to changing specialization are strong enough they seem to dominate the pattern of change so much that the results are insensitive to whether or not account is also taken to some other trade determinants viz. tariff changes, growth rate of demand and certain entry and exit barriers.

It is important to emphasize that our study has not penetrated the causality structure of the factor proportions theory. Therefore, it is impossible to say whether the change in Swedish specialization is attributable to external changes - for instance through factor endowment changes induced by the development of world market prices for products with different factor requirements or to internal causes. What can be said is that the wage and salary structure of important skill categories remained stable during very long periods of time. Thus, the direction of causality of the simple, comparative static models from exogenous changes in factor supplies to endogenous changes in factor prices and trade should be rejected as regards skilled labour categories. Models with factor-price rigidities ought to be more appropriate in the Swedish case. It should be noted that such models seem realistic even beyond the decade-long time periods studied here. That is important since the trade adjustment has been shown to require substantially more than ten years before losing strength.

The other major trade theory investigated in our study is the product cycle theory. That theory is analysed only at the commodity group level for a six year long period of changes in specialization. It cannot be separated from the factor proportions account with respect to either the technology variable used or the market growth rate variable. However, two other variables tried were specific to the product cycle theory. Indeed, the efforts to develop a method for a full-blown analysis of that theory seem to be successful from the fact that the various independent variables captured a good part of the technology and market characteristics that are assumed different for products in different stages of their life cycle *and* the fact that some of the variables proved to be significant ones.

However, we have explicitly left out from consideration the possible age of the product. The idea behind the new methodology was instead to identify the technology and market characteristics and their changes over time regardless of whether in fact they were associated with the differences in age between products. Although the method is not easily and totally applicable for all manufacturing trade some of its elements are. As far as we know it is the only method which combines explanatory variables of an inter-"industry" nature with variables of an intra-"industry" nature, for instance measures of product differentiation and its change and of technological heterogeneity of the "industry" or "market" and its change over time. The latter change could be looked upon as a variable measuring how fast the products included become standardized and the extent to which technologically different new products are developed and included in the "industry", or market.

The methodology used for the commodity group analysis did not allow us to compare directly the factor proportions account with the product cycle account. To some extent there is no need for such a discrimination. The former theory appears to be designed for long run changes in trade patterns according to both its inherent static properties and the results of our study. The product cycle theory should for similar reasons and in cross-sectional studies be more suitable for the medium run, i.e. the 5-10 year long perspective provided that the product "age" is interpreted only as an expression capturing differences between products with certain technology and market characteristics. Both theories hold better for changes in specialization than for the pattern in a given year. However, it appears as if those changes for fourteen industrial countries are not easily associable with the size and development level of the countries as originally suggested by Vernon and Hirsch.

10.4 Possible directions of further studies

The results of the present study suggest several possible directions for further studies of trade specialization patterns. One direction would be to analyse the Swedish case more in depth. Either one of three main tracks could be followed. For all three, the analysis should be extended to cover the whole manufacturing sector.

One analytical track would be to identify both long run and temporary determinants of the pattern of specialization in a given year. The present study was not successful on this point. That analysis should have to combine explanations from many trade theories whether it is focussed on inter-industry patterns or on intra-industry ones.

An alternative analytical track for Sweden would be to study the trend in specialization for the whole manufacturing sector, an analysis which should if possible capture a period longer than 10 years and in addition study more in depth possible factor (and product) market distortions[5]. Some of those distortions might be policy imposed. There may also be differences between the distortions as regards the time span during which they are effective.

A third track would be to use a multicountry instead

of a two-country framework. Such a study would further decrease the teoretical underpinning of the empirical analysis but could well be successful in terms of explanatory power according to earlier published factor proportions studies. It would seem appropriate from our results to direct the analysis of Swedish bilateral trade patterns toward different bilateral specialization patterns and their association with the time required for, for instance, market penetration of new exportables and the withdrawals of previous exportables during a period of changing comparative advantage.

Another direction of study is to analyse the trends in specialization for longer periods of time for other industrial countries, covering more than the engineering sector. Such an analysis would require good knowledge of the various industry classification systems used by the individual countries and of the definitions of the variables to be used in the construction of factor intensities. Therefore that task is suitable for cooperation among researchers of several countries. It appears to be potentially the most important one from the point of view of detecting trends of interest in policy oriented discussions of the changing international division of labour.

A third direction of further studies would be to develop the new methodology presented in our study for better analysis of the product life cycle theory and other theories associated with the concepts of intra-industry trade and product differentiation. From the point of view of forming effective medium-run industrial, regional and trade policies, the trade adjustment proposed by theories other than the factor proportions theory should be worthwhile to study further.

Footnotes

1. The disaggregation to the commodity group level meant an increase in the capital intensity (or a decrease in the technical personnel intensity) of the observations compared to the industry level.

2. Permanent plants " expansion entries - contraction exits = permanent firms (in a given industry).

3. Undertaken by the author for the Expert Group on Regional Studies, a government committee under the Ministry of Industry.

4. The exceptions are found for the entry and exit rates in their relationships with some independent variables (see chapter 7 and also Du Rietz [forthcoming]).

5. This is in fact the idea behind the above mentioned research project comparing Swedish regional and national specialization trends and including studies of the effects of regional policy on the structure of industry and entries of firm.

Appendix A

THE CHOICE OF TECHNOLOGY MEASURES AND ITS POSSIBLE IMPACT ON THE RESULTS

For two reasons the choice of variables in the main text to measure the relative factor use of the industries may lead to false conclusions. One reason is that practically all variables are proxy variables and as such cannot capture more than some of the theoretically desirable properties. Another reason is associated with the use of multifactor models. In such models, the factor intensity concept loses much of the neat theoretical properties it has in two factor models. It is no longer possible to derive theoretically a definition on how to measure the varying relative factor requirements of the products.

For both reasons an extensive sensitivity analysis of the industry results of chapter 4 was carried through. Chapter 5 constitutes one part of that analysis. The conclusions of the analysis presented here wholly support those of chapters 4 and 5. Therefore, the following presentation is kept short and is directed more towards indicating the alternatives that were possible.

A.1 Non-human capital measures

At the detailed industry level, Swedish industrial statistics offered three possible measures of the use of non-human capital. One measure had earlier been used by Carlsson & Sundström [1973], Carlsson & Ohlsson [1976], Lary [1968], Ohlsson [1969] and Åberg [1969, 1974], namely gross capital income defined as value added minus total wage sum. Theoretically and empirically that measure is not a satisfactory one for cross-sectional comparisons of *changes* in specialization if those changes can be assumed to be associated with changes in the comparative advantage in capital intensive production. Such a change was hypothetically discussed in Ohlsson [1969, 1973]. Consequently, the measure was ruled out without being tried.

Another possible measure used in earlier Swedish studies was the capacity of motive power of installed

Therefore, some efforts and costs were taken to construct better measures of that intensity utilizing the wage statistics of the two labour market organizations for the engineering industry, the Swedish Engineering Employers' Association, (VF), and Swedish Metalworkers' Union. Practically all plants of any importance are associated with VF. The wage statistics include apart from hourly wages also the number of worked hours per quarter of the year. Two cross-classifications of blue collar workers are used. First, A-, B- and C-categories are distinguished. The A -category is the skilled manual worker category. A worker is regarded as a skilled manual worker in the period if a) he/she has a three year long schooling in a given skill (in school or at the job) and b) the job he/she is actually carrying out requires those skills.

The second classification used distinguishes among types of skills. It separates about thirty types of skills of which 24 are so-called core skills for the engineering sector.

The first possible quarter for constructing skilled manual worker intensities by our industries was the fourth quarter of 1972. Having calculated the basic wage and hours-worked statistics by industry, many possible definitions of skilled manual workers could have been constructed. The most aggregate measure is that used in the main text, i.e. hours worked by the A-category treating the B- and C-categories as unskilled workers. A second definition used was the hours worked by the A-category within the so-called core skills. However, the regression results were almost identical for the skilled manual worker intensity based on core skills as for the one used in the main text. The correlation coefficient between the two measures is 0.997.

Both the variation in hourly wages for given skills and other information pointed to relatively substantial differences in skill levels within the A-category attributable to length of schooling or on the job training. Therefore, a measure reflecting those differences was constructed. Either the wage per hour or a ratio between the wage per hour for skilled workers and that for unskilled workers could have been used. The latter measure was chosen due to evidence of a regional variation in absolute wages which did not seem related to a corresponding variation in skill levels.

A.3 Certain correlation and regression results for alternative factor intensity measures

The following correlation coefficients indicate how sensitive the conclusions for the industry patterns of engineering trade specialization are with respect to the alternative capital/labour measures (34 observations):

	Horsepower/ employee	Electricity consumption/ employee
Net export ratio 1970	0.074	-0.274
Home market share 1970	0.393[a]	0.431[a]
Change in net exports ratio 1960-1970	-0.353[a]	-0.441[a]
Change in home market share 1960-1970	0.156	0.278

[a]significant at the 5% level (one-tail test)

The net export ratio was in 1970 not correlated with either capital intensity measure. Both measures lead to the rejection of the hypothesis that Sweden had a comparative advantage in capital intensive production. Both measures give support to this hypothesis if we instead use the home market share. Furthermore, both measures indicate that Sweden diminished its specialization in capital intensive industries according to the changes in the net exports ratio but not according to the changing home market shares. The results are strikingly similar to those of chapters 4 and 6.

Table A:1 presents regressions on the net exports ratio and on the home market share in 1970 defining the factor intensities with unskilled labour in the denominator. To derive the latter measure the sum of office-clerks and unskilled manual workers (in estimated man-years) was estimated. With one exception the results are again strikingly similar to those of chapter 4. The exception is that L_Y/L_U does not receive a significant negative coefficient in the regression on $(H/C)_{70}$ although L_Y/L_A was significant (at the 10% level).

Table A:2 presents regression results when the factor intensities are defined with value added in the denominators. With no exception the results are the

Table A:1

Regressions on the 1970 specialization utilizing un- or low-skilled workers in the denominators of the factor intensities

Dependent variable	Constant	Regression coefficients (with standard errors) for K/L_U	L_T/L_U	L_Y/L_U	t_{70}	R^2	F-value (degrees of freedom)
$(\frac{X-M}{X+M})_{70}$	-0.159	-0.002	-0.423	0.537[c]	3.641[a]	0.214	1.767
		(0.006)	(0.406)	(0.227)	(2.563)		(4:26)
$(\frac{H}{C})_{70}$	0.784	0.007[b]	-0.565[c]	-0.078	-2.265[a]	0.437	5.045[c]
		(0.004)	(0.231)	(0.129)	(1.458)		(4;26)

[a] Significant at the 10% level (one-tail test)
[b] " " " 5% " " " "
[c] " " " 2.5% " " " "

Legend (complementing those in chapter 4):

L_U = Number of office-clerks + man-years worked by unskilled manual workers (B- and C-categories).

Table A:2

Regressions on the 1970 specialization utilizing value added in the denominators of the factor intensities

Dependent variable	Constant	Regression coefficients (with standard errors) for K/v	L_T/v	L_Y/v	t_{70}	R^2	F-value (degrees of freedom)
$(\frac{X-M}{X+M})_{70}$	-0.154	-0.050	-4.066	7.702[c]	3.226	0.206	1.688
		(0.108)	(5.216)	(3.216)	(2.540)		(4;26)
$(\frac{H}{C})_{70}$	0.762	0.118[b]	-5.357[b]	-2.529[a]	-2.190[a]	0.392	4.190[c]
		(0.064)	(3.069)	(1.893)	(1.494)		(4;26)

[a] Significant at the 10% level (one-tail test)
[b] " " " 5% " " " "
[c] " " " 2.5% " " " "

Legend (complementing those in chapter 4)

v = value added.

same as in chapter 4 with respect to signs, significance and explanatory power.

The conclusion of the sensitivity analysis presented here is that the results of chapter 4 are not especially sensitive to changes in the empirical definitions of the factor intensities. Combined with the results of chapter 5, this appendix demonstrates that the low explanatory power of the factor intensities in regressions against the net export ratio does not seem to be associated with errors of measurement in the independent variables. Instead the explanation should be economic in kind, for instance that Sweden was on the move from a capital abundant economy towards a human capital abundant one without having reached a point where capital had become relatively scarce or human capital relatively abundant.

Footnote

1. The correlation coefficient between the wage per hour and the skilled manual worker intensity is 0.47.

Appendix B

THE TON PRICE AS A MEASURE OF TECHNOLOGICAL DIFFERENCES BETWEEN ENGINEERING PRODUCTS

Comparative country studies of engineering trade specialization have been difficult for two reasons. One reason is the lack of uniform and detailed industry breakdowns. The size and heterogeneity of the engineering industry demands a substantially more detailed classification than the traditional four or five major engineering industries[1].

International trade statistics published by various international organizations offer detailed commodity group breakdowns. However, these statistical sources did not appear to provide any measures of the factor use of different commodity groups. This limitation constituted the second reason preventing ambitious comparative country studies.

Up until 1971 it was impossible to carry through meaningful cross-sectional analysis using Swedish industrial statistics since they included technology data on only five major engineering sectors. That was the background of and reason for the development of the ton price methodology used in chapters 5, 8 and 9 of the present study. Earlier publications presenting parts of that methodology are Ohlsson [1973, 1974, 1976]. A critical review is Wijkman [1974]. That the price per unit of weight measures qualitative differences between products is not an idea original to our study. Paues [1958] demonstrated systematic country differences in such prices for textile trade. Burenstam Linder, Grubel and Keesing have all explored the same idea[2]. Hufbauer [1970] and Hufbauer & O'Neill [1972] used ton prices on the same basis. Kravis & Lipsey [1971a and b] gave evidence that the export ton prices of investment goods could hardly be looked upon as commodity prices.

Despite these previous efforts no solid theoretical or empirical basis has been established as to what the ton price actually reflects. As regards engineering products all products are sold by piece and not per unit of weight. The following section aims at exploring the relationships between the metric ton price of engineering products and factor intensities.

B.1 The ton price as a technology proxy

Traditionally, engineering products are manufactures of various metals which are mechanically or otherwise physically transformed. Suppose that the production process of commodity j transforms a homogeneous metal bought in tons into pieces of commodity j. Since $P_{tj}O_{tj} = P_jO_j$ (where P_t = price per ton, O_t = quantity in tons, P = price per piece of the commodity and O = number of pieces) the following identity holds for product/sector j:

$$P_{tj} - Z \equiv w\ell_j + rk_j \qquad j = 1,\ldots,n,$$

where Z = the ton price of the metal
w = the wage rate
ℓ_j = labour input per ton of the product
r = the price of capital
k_j = capital input per ton of the product.

The expression assumes no waste of the raw material and identical prices for all sectors for the raw material, labour and capital inputs.

Looking at successive steps of the processing of the raw material it is certain that the price of the output of the *integrated* process must rise with each step. However, the value added price of each step does not have to increase along a given manufacturing line. It is also true that if two products use the same raw material but one uses more of both labour and capital per ton of input that product will have the higher output price per ton.

For other products and at given factor prices, the differences in output ton prices are attributable to the different factor combination of the products and relative factor prices. Hence, ton prices can only be explained, theoretically and empirically, if we can provide an explanation as to why the producers choose different factor intensities at the same factor prices, i.e. can we explain the choice of technology?

It is obvious that a producer uses lower and cheaper skills unless the more expensive, higher skills pay in terms of a higher productivity or a higher output price. Comparing two products with different skill proportions the one with the higher skill intensity is therefore assumed to have the higher ton price. Hence, we exclude the possibility that the use of higher skills only leads to a pure productivity effect. Usually higher skills are used for other tasks which eventually leads to higher priced products. For similar reasons a more mechanized technology is chosen if the saving on labour costs more than compensates the increase in cost of capital. Comparing two products differing only with respect to their capital/labour ratio, the one with the lower capital/labour ratio should receive the higher ton price for either of two reasons:

1. The product with the higher K/L has longer production runs (or its components have longer runs). Therefore the processes were easier to mechanize.
2. The product with the higher K/L (or its components) is less complex with respect to the number or character of production steps. Therefore the processes were easier to mechanize.

According to our discussion the ton price of a process should be positively related to the human capital intensiveness and negatively to the non-human intensity.

B.2 The test method

It was impossible to test how the value added ton price of a process was related to the factor intensities of that process. For one reason the published output values and quantities are gross rather than net values. Secondly, only about half of the commodity groups have quantities measured in metric tons. Thirdly, the industries do not use only one homogeneous metal[3].

These deficiencies of our regression method may or may not disqualify it as a formal test of the assumed relationships. In any event, the method is as follows. The metric ton prices of exports and imports are respectively used as dependent variables. There is a risk that the relative distribution of production across the commodity groups within a given industry deviates from the corresponding distributions for exports and imports. For a given industry the ton prices of exports and imports were estimated according to the formula

$$\sum_{i=1}^{s} y_i / \sum_{i=1}^{s} q_{yi}, \qquad (i = 1,\ldots,s \text{ products of the industry})$$

where y_i = value of exports (or imports) of commodity j

and q_{yi} = quantity in metric tons of exports (or imports) of that commodity.

The lowest aggregation level according to the Brussels nomenclature in Swedish trade statistics was used.

The estimated ton prices (of the integrated "product")

were regressed on the *direct* factor intensities of the value added of the industries as discussed in the following.

B.3 Regression results

Ohlsson [1974] reported regression results for 40 industries and the year 1967. Apart from the capital intensity, three skill intensities were included as independent variables, namely the technical personnel intensity, the skilled manual worker intensity measured by the wage per hour for manual workers and the sales personnel intensity. A fifth independent variable was the proportion of workers employed in plants with less than 200 workers. That "small plant intensity" was assumed to capture the influence of interindustry differences in scale economies and optimum scale levels (cf Ohlsson [1974]).

The regression analysis to be presented here deviates from Ohlsson [1974] in several respects. First, the engineering sectors included are the 31 industries of chapters 4 and 6 of the study and the year covered is 1970. Secondly, the skilled manual worker intensity is measured by L_Y/L_A discussed elsewhere in the study rather than w_A, the wage per hour for manual workers. Thirdly, the capital intensity has also been measured by the electricity consumption/employee. Since that variable generated lower R^2 values results are only reported for the horsepower/employee variable. Fourthly, a number of functional forms have been tried for especially the ton price relationships with K/L and L_T/L. Table B:1 includes only some of those regressions, namely those which obtained the highest R^2 values. Thus L/K gave a higher explanatory power than K/L. The ton price relationship with L_T/L appeared to be polynomial which motivated us to include both L_T/L and $(L_T/L)^2$ in the regressions. Generally, the variables included explained a larger proportion of the variance in P_X, the export ton price, than in P_M, the import ton price. Between 87 and 89% of the former variable could thus be explained but at most only 73% of the latter.

Table B:1 gives strong evidence that the ton price receives higher values the more labour per unit of capital (L/K) is used. It is likewise true that the ton price is higher the more technical personnel intensive the

Table B:1

Regressions on the export and import ton prices of 21 engineering industries in 1970

Regression no.	Dependent variable	Constant	Regression coefficients (with standard errors) for: L/K	L_T/L	$(L_T/L)^2$	L_Y/L_A	L_F/L	S	R^2	F-value (degrees of freedom)
1	P_X	-23.2	110.1[c] (11.3)	107.9[c] (33.9)		13.3 (10.7)			0.872	61.18[c] (3;27)
2	P_M	-69.2	134.7[c] (65.3)	544.0[c] (195.8)		7.4 (61.9)			0.478	8.23[c] (3;27)
3	P_X	-28.0	105.7[c] (12.3)	149.4[c] (53.7)		5.8 (13.4)	-85.1 (189.9)	12.0 (11.9)	0.877	35.69[c] (5;25)
4	P_M	-85.8	106.6[a] (69.3)	780.3[c] (303.9)		-31.2 (75.7)	-1368.2 (1074.2)	67.2 (67.4)	0.521	5.44[b] (5;25)
5	P_X	-12.3	111.8[c] (10.7)	-92.3 (102.5)	618.9[c] (301.0)	16.0[a] (10.2)			0.886	54.42[c] (4;26)
6	P_M	45.2	152.6[c] (49.1)	-1550.9[c] (470.5)	6476.3[c] (1381.8)	35.4 (46.8)			0.717	16.46[c] (4;26)
7	P_X	-15.3	109.7[c] (12.0)	-58.1 (131.4)	569.8[b] (331.7)	12.4 (13.5)	-30.0 (185.7)	5.3 (12.1)	0.891	32.56[c] (6;24)
8	P_M	55.3	152.3[c] (54.4)	-1547.2[c] (594.8)	6389.4[c] (1500.9)	43.3 (60.9)	-749.9 (840.2)	-7.8 (54.8)	0.727	10.66[c] (6;24)
9	P_X	-17.0	109.6[c] (10.4)		361.3[c] (93.6)	13.6[a] (9.1)			0.886	70.12[c] (3;27)

[a] Significant at the 10% level (one-tail-test)
[b] " " " 5% " " " "
[c] " " " 2.5% " " " "

Legend (complementing those of chapter 4):

P_X = The metric ton price of Swedish exports
P_M = " " " " " " imports
L_F/L = Sales personnel per employee.
S = Proportion of manual workers employed in plants with less than 200 manual workers.

technology is. According to regressions 5 and 6, the skilled manual worker intensity has also a weak positive influence on the ton price. Hence, the "rest of the employees" intensity $(L-L_T-L_Y)/L$ should have a negative influence on the ton price.

In contrast to the results reported in Ohlsson [1974] neither the sales personnel intensity nor the "small plant" intensity receive significant coefficients.

In conclusion, the ton price is found to vary negatively with the capital intensity and positively with the technical personnel and skilled manual worker intensities. It is the former two independent variables that generate the very high explanatory values of the regressions. Those results establish the empirical basis of the use of the metric ton price as a proxy variable for those two intensities in chapters 5, 8 and 9. It is expected that other countries with other relative factor prices should have other magnitudes of the corresponding regression coeffieients to those of table B:1. However, the signs should be the same due to irreversible factor intensities and equalized commodity prices.

Footnotes

1. The fabricated metal product, non-electrical machinery, electro-technical, the transport equipment and sometimes the shipyard industries.
2. Their analyses have not been published. Grubel [1967] mentions his efforts in a footnote while the other two have mentioned their work in private letters and discussions.
3. According to qualitative information it can be presumed that the use of more expensive metals combines with a lower use of capital and a higher one of technical personnel and skilled manual workers.

Appendix C

Basic Data

Table C:1

Exports, imports, gross output and domestic consumption by industry 1960 and 1970 used in chapters 4 and 6.

Thousands of Swedish crowns

	1960				1970			
SNI-no.	Exports	Imports	Gross output	Domestic consumption	Exports	Imports	Gross output	Domestic consumption
3811	198 742	51 164	328 173	180 595	502 223	178 771	976 956	653 504
3813	45 009	41 293	742 295	738 579	177 752	139 985	2175 186	2137 419
38191	7 643	9 251	162 585	164 193	12 675	37 339	371 455	369 119
38192	55 262	38 113	368 881	351 732	84 124	136 473	765 932	818 281
38193	32 481	31 321	204 598	203 438	83 290	110 125	402 174	429 009
38194	217 377	134 745	574 974	492 342	373 999	267 971	1174 929	1068 901
38195	46 115	21 094	148 904	123 883	76 015	54 098	221 160	199 243
38199	137 588	308 706	827 626	998 744	689 804	948 653	1920 253	2179 102
3821	51 200	28 266	152 428	129 494	138 777	118 055	430 555	409 833
3822	103 242	66 967	313 981	277 706	313 287	267 834	942 048	896 595
38231	134 173	161 775	300 470	328 072	453 488	468 646	723 214	738 372
38232	41 565	18 612	80 326	57 373	128 437	46 543	234 069	152 175
38241	152 060	63 510	213 460	124 910	322 420	119 480	372 870	169 920
38242	77 359	75 948	230 505	229 094	324 647	204 016	658 970	538 339
38249	270 008	210 554	451 451	391 997	1051 610	660 807	1507 008	1116 205
38251	15 547	56 635	50 184	91 272	233 534	310 706	331 305	408 477
38259	130 865	46 024	159 057	74 216	641 766	405 119	729 835	493 188
38291	56 425	49 394	312 081	305 050	264 702	163 222	642 557	541 077
382991	89 095	91 026	279 844	285 369	612 983	384 533	1068 597	840 147
382992	42 147	51 125	112 418	121 396	236 049	193 579	336 292	293 822
382993	213 921	79 636	365 416	231 131	422 629	310 550	663 504	551 425
382999	314 734	256 402	727 090	668 758	923 365	836 815	2007 161	1920 611
3831	149 632	113 543	463 953	427 864	427 485	375 993	941 095	889 603
3832	238 730	420 750	898 700	1081 530	1384 420	1164 440	2675 320	2455 350
3833	45 931	31 817	158 766	144 652	113 792	131 539	435 239	452 986
38391	24 800	29 550	304 880	309 620	120 820	127 010	842 660	848 840
38392	24 283	9 884	76 539	62 140	66 500	39 483	226 929	199 912

Cont.

	1960				1970			
SNI-no.	Exports	Imports	Gross output	Domestic consumption	Exports	Imports	Gross output	Domestic consumption
38393	2 334	25 756	44 588	68 010	8 238	90 260	68 300	150 322
38399	28 055	217 443	231 176	420 564	155 060	730 183	647 494	1222 617
38411	920 739	217 626	1289 428	586 315	1053 509	340 023	1317 390	603 904
38412	9 822	15 400	19 885	25 463	41 605	61 234	161 084	180 713
38413	115 730	48 880	196 240	129 380	302 640	92 790	605 230	395 370
38421	20 014	12 371	130 929	123 286	77 379	41 696	197 683	162 000
38431	568 691	712 273	1336 873	1480 455	2532 178	1180 630	3963 437	2611 889
38432	116 800	339 508	814 835	1037 543	746 162	1160 108	2283 811	2697 757
3844	15 091	33 528	145 884	164 321	29 556	105 771	132 299	208 514
3845	26 039	269 078	381 681	624 720	148 804	367 405	676 476	865 077
3849	6 101	10 175	34 446	38 520	27 022	9 726	67 508	50 212
3851	64 878	174 122	167 554	276 798	378 075	515 355	600 436	737 716

Note: Domestic consumption is defined as gross output (*O*) minus exports, fob (*X*) plus imports, cif (*M*). The latter three variables are aggregated from commodity group statistics at the lowest possible level of aggregation utilizing a key between the SNI codes and the SITC and Brussels nomenclatures. This key was kindly provided by the Swedish Central Bureau of Statistics (SCB). The names of the industries are provided in the main tables of chapters 1, 4 and 6.

Sources: SOS, Industri 1960, del 2 1970, and *SOS*, Handel, del 1 1960, *SOS*, Utrikeshandel, del 1 1970.

Table C:2

Other basic data of chapters 2 and 4

SNI-no.	Tariff in % 1960	1970	Change in tariff 1960-70	Growth of domestic consumption 1960-70 in %
3811	7.8	4.6	-3.2	262
3813	8.5	2.1	-6.4	189
38191	7.6	1.4	-6.2	141
38192	7.3	2.5	-4.8	133
38193	7.4	3.8	-3.7	111
38194	4.4	3.3	-1.2	117
38195	9.1	3.2	-5.9	61
38199	4.9	2.4	-2.5	118
3821	9.6	1.2	-8.4	217
3822	9.6	2.9	-6.7	223
38231	8.3	3.0	-5.3	125
38232	9.0	5.0	-4.0	165
38241	8.9	2.6	-6.2	36
38242	9.7	3.7	-6.0	135
38249	7.7	3.1	-4.6	185
38251	9.8	6.2	-3.7	348
38259	9.6	5.9	-3.7	565
38291	9.6	4.8	-4.8	77
382991	9.8	3.7	-6.1	194
382992	9.6	4.2	-5.4	142
382993	9.7	4.2	-5.5	139
382999	8.5	3.9	-4.6	187
3831	9.7	3.4	-6.4	108
3832	14.4	5.2	-9.2	127
3833	9.6	3.2	-6.3	213
38391	8.7	5.6	-3.1	174
38392	8.4	2.4	-6.1	222

Cont.

(C:2 cont.)

SNI-no.	Tariff in %		Change in tariff 1960-70	Growth of domestic consumption 1960-70 in %
	1960	1970		
38399	9.9	5.9	-4.1	191
38413	8.2	3.2	-5.0	206
38421	8.2	3.7	-4.5	31
38431	14.5	13.3	-1.2	76
38432	11.5	5.0	-6.5	160
3844	13.0	3.2	-9.8	27
3851	8.6	4.5	-4.0	167
The Swedish engineering industry				
Unweighted	9.1	4.0	-5.1	
Weighted	10.3	5.0	-5.2	

Note: The percentage tariff of an industry is calculated as total tariff revenues in per cent of total Swedish imports of commodity groups classified to this industry. Growth of domestic consumption (C) is measured as $100 \cdot (C_{1970} - C_{1960})/C_{1960}$.

Sources: See table C:1.

Table C:3

Basic data for chapters 8 and 9 on measures of the Swedish commodity group specialization 1964-1970

SITC no. (1)	Net export ratio 1970 (2)	Change 1964-70 in net export ratios (3)	World export share 1970 (4)	Change 1964-70 in world export shares (5)	Home market share (1970) (6)	Change 1960-70 in home market shares (7)
6911	0.1504	0.3897	0.0291	0.0204	0.9467	-0.0273
6912	-0.2562	-0.5457	0.0148	-0.0029	0.9045	0.0624
6913	1.0000	0.7059	0.0081	-0.0395	1.0000	0.0
6921	0.1637	-0.1307	0.0180	0.0034	0.9455	-0.0280
6922	-0.4222	-0.3529	0.0160	-0.0087	0.9207	-0.0285
6923	0.0753	-0.0050	0.0288	-0.0092	0.8391	0.1249
6931	-0.6528	-0.4562	0.0101	-0.0070	0.5530	-0.2811
6932	-1.0000	-0.2597	0.0	-0.0003	0.5131	-0.4869
6933	0.0790	-0.1443	0.0418	-0.0059	0.8061	-0.0236
6934	-0.6984	0.2728	0.0176	0.0159	0.0252	0.0185
6941	0.0768	-0.3422	0.0318	0.0078	0.8128	-0.1368
6942	-0.1710	-0.1644	0.0280	-0.0113	0.7346	-0.0877
6951	0.3946	-0.0760	0.0432	0.0070	0.0	-0.9409
6952	0.5430	-0.0608	0.0992	-0.0101	0.7464	-0.0169
6960	-0.2483	-0.0273	0.0131	-0.0054	0.4905	-0.1342
6971	0.5467	0.0084	0.0293	-0.0135	0.8227	-0.0378
6972	0.1960	-0.1943	0.0297	-0.0227	0.8023	-0.1021
6979	-0.3328	0.1865	0.0168	0.0022	0.5330	-0.2672
6981	-0.5277	-0.1957	0.0178	-0.0068	0.6106	-0.1025
6982	0.2688	0.2243	0.0443	0.0099	0.9256	-0.0505
6983	0.0869	0.1866	0.0445	0.0081	0.6363	-0.0827
6984	-0.5082	0.4017	0.0180	0.0130	0.1043	0.1340
6985	-0.4059	0.1781	0.0145	0.0016	0.1398	-0.2570
6986	-0.6429	-0.0146	0.0146	0.0008	0.7636	-0.0801
6988	-0.2268	-0.1902	0.0312	-0.0121	0.6981	-0.0949
6989	-0.0860	0.2310	0.0373	0.0171	0.8070	-0.0592
8121	0.6843	-0.0526	0.0781	-0.0342	0.9314	-0.0479
8122	0.6116	-0.0683	0.1558	0.0333	0.8537	-0.0292
8123	0.8840	0.0079	0.1751	0.0396	0.9759	-0.0043
8124	-0.1251	-0.0870	0.0330	0.0047	0.8066	-0.0884
8611	-0.9504	-0.0231	0.0015	-0.0009	0.1343	-0.0472
8612	-0.7747	0.1429	0.0043	0.0023	0.1063	0.0439
8613	-0.7139	0.0946	0.0044	0.0001	0.3819	0.3634
8614	-0.3205	0.1217	0.0237	0.0079	0.0792	0.0553
8615	-0.8973	0.0120	0.0017	-0.0003	-0.0572	-0.1329
8616	-0.4195	0.2817	0.0135	0.0070	-0.0885	-0.1507
8617	0.0445	0.2412	0.0378	0.0123	0.0905	-0.1022
8618	-0.6373	-0.1744	0.0157	-0.0123	0.1344	0.1077
8619	-0.2318	0.1743	0.0222	0.0084	0.3738	-0.0720
7111	0.2736	0.2711	0.0280	0.0132	0.9107	0.0933
7112	-0.5685	-0.8090	0.0064	-0.0105	0.7694	-0.1460
7113	0.3959	-0.4942	0.0720	0.0089	0.7815	-0.0199
7115	-0.0858	0.0345	0.0262	-0.0039	0.5562	-0.0613
7116	0.0302	0.8405	0.0265	0.0223	0.4986	-0.5203
7117	0.3382	-0.5072	0.0099	-0.0012	0.6967	-0.2679
7118	0.1295	-0.1715	0.0310	-0.0158	0.6235	-0.2878
7120	0.2307	0.1145	0.0300	0.0048	0.7134	0.0911
7141	0.1870	-0.0299	0.0262	-0.0086	0.3519	-0.1176
7142	0.2454	-0.1248	0.0795	-0.0555	0.2632	-0.3721
7143	0.1410	0.2179	0.0124	-0.0130	0.2782	-0.2414
7149	-0.1979	0.1214	0.0228	-0.0029	0.0669	-0.1021
7151	-0.1539	0.0239	0.0219	0.0019	0.2610	-0.1126

cont.

(C:3 cont.)

SITC no. (1)	Net export ratio 1970 (2)	Change 1964-70 in net export ratios (3)	World export share 1970 (4)	Change 1964-70 in world export shares (5)	Home market share (1970) (6)	Change 1960-70 in home market shares (7)
7152	-0.0350	0.0359	0.0365	0.0089	0.6731	0.0389
7171	0.0278	0.3081	0.0095	0.0023	0.0883	-0.0905
7172	-0.2024	-0.3532	0.0038	-0.0198	0.1617	-0.2514
7173	0.2078	0.1041	0.0217	-0.0012	0.6777	0.0062
7181	0.3460	0.0607	0.0967	0.0177	0.2566	-0.1760
7182	-0.0380	0.1134	0.0290	0.0010	0.5036	-0.0144
7183	-0.0139	0.1537	0.0189	0.0085	0.4971	-0.1849
7184	0.2143	1.0172	0.0231	0.0021	0.5712	-0.0145
7185	0.4932	0.9990	0.0597	0.0028	0.6140	-0.2068
7191	0.2592	-0.1484	0.0456	-0.0082	0.6376	-0.1347
7192	0.0868	-0.0072	0.0513	-0.0078	0.2729	-0.2367
7193	0.2283	0.1840	0.0563	0.0256	0.5475	-0.1357
7194	0.4331	-0.2310	0.0580	-0.0553	0.1968	-0.5835
7195	0.4627	-0.0373	0.0707	-0.0214	0.4603	-0.1453
7196	0.0901	0.1430	0.0374	0.0093	0.5094	-0.0591
7197	0.5962	-0.2335	0.1014	-0.0652	0.5523	-0.2929
7198	-0.0065	-0.0762	0.0269	-0.0006	0.5526	-0.0258
7199	-0.3791	-0.0192	0.0229	-0.0036	0.5306	-0.0701
7221	-0.0666	-0.3326	0.0390	-0.0229	0.5772	-0.1496
7222	-0.4353	-0.1274	0.0192	-0.0102	0.4762	-0.1057
7231	-0.0976	0.1699	0.0302	0.0142	0.8457	-0.0542
7232	-0.4312	0.0748	0.0124	0.0036	0.6567	-0.0025
7241	-0.2396	0.1682	0.0333	0.0229	0.4017	-0.1946
7242	-0.8074	-0.1282	0.0022	-0.0024	0.2314	-0.3300
7249	0.4793	0.1542	0.0732	0.0230	0.6948	-0.0340
7250	0.4532	0.2654	0.0883	0.0315	0.7006	-0.1330
7250	-0.7796	-0.0825	0.0049	-0.0053	0.6086	-0.1183
7250	0.3010	0.1784	0.0497	-0.0070	0.7274	0.0404
7250	-0.2604	-0.1580	0.0320	-0.0015	0.7324	-0.1696
7261	0.5918	0.0745	0.1034	-0.0393	0.3894	0.4117
7262	0.2380	-0.0674	0.0376	-0.0252	0.0085	-0.3893
7291	0.2547	-0.1146	0.0468	-0.0083	0.7871	-0.0540
7293	-0.8367	0.0291	0.0030	0.0003	-0.0062	-0.0387
7294	-0.9343	-0.0173	0.0024	-0.0015	0.1119	-0.0436
7295	-0.2997	0.3317	0.0151	0.0049	0.0380	-0.0685
7296	-0.5337	0.0239	0.0118	0.0001	0.0745	-0.2668
7299	-0.0409	0.1282	0.0321	0.0088	0.5358	-0.0702
7312	0.0900	-0.9100	0.0399	0.0393	-0.0156	-1.0156
7313	0.9258	0.0154	0.0950	0.0799	0.8816	0.2004
7314	-0.9294	-1.9294	0.0006	-0.0049	0.0013	-0.9987
7315	1.0000	1.5000	0.0003	0.0003	1.0000	0.0
7316	0.7150	-0.0602	0.0484	0.0060	0.9224	-0.0397
7317	-0.1041	-0.1679	0.0337	0.0023	0.4564	-0.3306
7321	0.2418	0.5840	0.0309	0.0038	0.5031	0.0691
7322	-0.6683	0.2135	0.0071	0.0047	0.8075	-0.0459
7323	0.7652	0.1496	0.0572	-0.0018	0.7294	-0.0248
7324	-0.4443	-0.3618	0.0042	-0.0014	0.1846	0.1220
7325	0.9795	-0.0205	0.0103	0.0098	0.0	0.0
7327	0.8150	-0.1092	0.0439	-0.0247	0.9332	-0.0280
7328	-0.2253	0.2251	0.0171	0.0015	0.6131	-0.0389
7329	-0.3296	0.2596	0.0063	0.0022	0.4874	-0.3151
7331	-0.7386	-0.4225	0.0074	-0.0049	0.5524	-0.3851
7333	0.0354	0.2084	0.0428	0.0070	0.7292	-0.0989
7334	-0.8038	-0.5181	0.0442	0.0004	-0.0229	-0.3563

Sources: OECD, Series C, Commodity trade statistics; Exports and Imports, 1964 and 1970; *SOS*, Handel, del 2 1960; *SOS*, Utrikeshandel, del 2 1964 and 1970, and *SOS*, Industri 1960 and del 2, 1970.

Table C:4

Basic data for chapter 8. Independent variables in regressions on the Swedish commodity group specialization 1970

SNI no.	X_W $1000	C 1000 Skr	P_E $1000/ metric ton	P_S/P_E	h	t_{70}	r
(1)	(2)	(3)	(4)	(5)	(6)	(7)	(8)
6911	744574.	300080.	0.444	0.646	0.305	0.0212	0.4255
6912	86632.	22683.	2.198	0.924	0.169	0.0111	0.5850
6913	740.	263.	0.719	0.758	0.381	0.0	0.4754
6921	124390.	29472.	0.604	1.733	0.396	0.0187	0.6169
6922	161537.	80118.	0.650	0.648	0.501	0.0156	0.3369
6923	63197.	9731.	0.656	1.587	0.409	0.0409	0.7331
6931	260296.	28097.	0.633	0.991	0.349	0.0196	0.8051
6932	24104.	819.	0.217	1.783	1.922	0.0727	0.9342
6933	150770.	27739.	0.428	4.813	1.556	0.0320	0.6892
6934	5862.	595.	0.391	8.724	1.019	0.0310	0.8157
6941	101993.	14832.	0.420	1.917	0.982	0.0353	0.7461
6942	448721.	66927.	0.976	1.043	0.965	0.0388	0.7404
6951	40927.	-896.	0.956	1.204	0.250	0.0156	1.0448
6952	1075662.	124654.	4.207	1.413	0.499	0.0456	0.7923
6960	326914.	13976.	7.343	1.219	0.313	0.0430	0.9180
6971	156742.	7603.	1.168	2.183	0.491	0.0223	0.9075
6972	225282.	22789.	2.175	1.296	0.289	0.0360	0.8163
6979	45366.	3264.	2.753	1.113	0.380	0.0236	0.8658
6981	376211.	55659.	2.041	0.784	0.205	0.0398	0.7422
6982	17141.	5889.	0.907	1.136	0.315	0.0114	0.4886
6983	145414.	14959.	0.871	0.504	0.931	0.0412	0.8134
6984	6659.	411.	0.419	1.656	0.498	0.0109	0.8838
6985	118765.	4729.	3.848	0.740	0.357	0.0359	0.9234
6986	75649.	21487.	0.764	2.500	1.111	0.0234	0.5576
6988	217179.	35630.	1.113	0.611	0.665	0.0231	0.7181
6989	669551.	153583.	1.062	1.654	0.316	0.0269	0.6268
8121	211966.	45189.	0.584	1.192	0.183	0.0255	0.6485
8122	63198.	16218.	0.504	2.260	0.674	0.0114	0.5916
8123	69451.	31029.	0.805	2.517	0.648	0.0401	0.3824
8124	308884.	67724.	2.498	1.133	0.351	0.0484	0.6403
8611	155961.	10759.	59.358	0.570	0.588	0.0460	0.8709
8612	157625.	5957.	25.506	2.409	0.580	0.0505	0.9272
8613	183931.	7832.	30.775	0.691	0.686	0.0279	0.9183
8614	497960.	24887.	13.570	1.953	0.701	0.0224	0.9048
8615	154931.	4701.	21.346	0.741	0.509	0.0505	0.9411
8616	311936.	9446.	6.552	0.990	0.241	0.0615	0.9412
8617	354454.	13470.	14.558	0.889	0.503	0.0495	0.9268
8618	97379.	7978.	8.640	2.903	0.592	0.0633	0.8486
8619	1299534.	73800.	10.226	1.247	0.418	0.0473	0.8925
7111	221368.	39582.	1.293	1.188	0.262	0.0218	0.6966
7112	89013.	8936.	1.793	1.639	0.323	0.0204	0.8175
7113	224107.	32120.	3.857	1.278	0.377	0.0060	0.7493
7115	2656368.	186375.	2.493	1.273	0.172	0.0393	0.8689
7116	176609.	8124.	7.356	0.146	1.233	0.0036	0.9120
7117	74430.	1197.	5.218	8.792	1.028	0.0358	0.9683
7118	150897.	9587.	3.321	0.800	0.346	0.0335	0.8805
7120	2153815.	141045.	1.287	1.485	0.215	0.0246	0.8771
7141	347755.	9611.	7.853	0.819	0.589	0.0695	0.9462
7142	1391019.	90944.	17.101	0.778	0.253	0.0658	0.8773
7143	737759.	9542.	34.173	1.006	0.537	0.0256	0.9745
7149	1645432.	59970.	17.322	0.869	0.754	0.0572	0.9297
7151	1977104.	79787.	3.204	0.855	0.342	0.0258	0.9224
7152	444168.	53190.	0.769	1.377	0.782	0.0345	0.7861

cont.

(C:4 cont.)

(1)	(2)	(3)	(4)	(5)	(6)	(7)	(8)
7171	2291512.	22572.	3.831	0.884	0.229	0.0146	0.9805
7172	89121.	602.	3.876	0.534	0.312	0.0158	0.9866
7173	426622.	18798.	5.321	1.125	0.437	0.0406	0.9156
7181	594303.	37549.	2.904	0.924	0.265	0.0273	0.8811
7182	736109.	46416.	4.203	0.999	0.257	0.0064	0.8814
7183	352038.	13606.	2.853	1.025	0.220	0.0384	0.9256
7184	1926349.	67168.	1.545	0.970	0.240	0.0412	0.9326
7185	608043.	31894.	1.901	1.475	0.409	0.0338	0.9003
7191	1859424.	137673.	2.148	1.364	0.277	0.0383	0.8621
7192	2140770.	127038.	3.158	1.180	0.268	0.0387	0.8880
7193	2103637.	164398.	1.561	1.085	0.238	0.0374	0.8550
7194	80969.	2313.	2.316	0.612	0.299	0.0398	0.9444
7195	1042556.	50153.	3.447	1.659	0.327	0.0492	0.9082
7196	1058237.	67414.	3.927	1.406	0.305	0.0520	0.8802
7197	633290.	36294.	3.169	0.887	0.486	0.0533	0.8916
7198	1867556.	113870.	3.306	0.983	0.155	0.0370	0.8851
7199	2043530.	221167.	2.663	0.594	0.244	0.0350	0.8047
7221	1555569.	163964.	2.376	0.920	0.260	0.0310	0.8093
7222	1707547.	158793.	6.525	0.711	0.307	0.0602	0.8298
7231	709660.	169110.	1.520	0.978	0.314	0.0532	0.6151
7232	126984.	11535.	0.803	1.166	0.780	0.0682	0.8335
7241	789018.	71703.	5.228	1.369	0.192	0.0498	0.8334
7242	1003678.	27465.	8.393	1.283	0.240	0.0889	0.9467
7249	2697765.	227667.	10.926	1.016	0.565	0.0577	0.8444
7250	382361.	42443.	1.275	1.193	0.279	0.0337	0.8002
7250	300333.	30407.	1.242	1.605	0.352	0.0655	0.8161
7250	246634.	24154.	3.154	1.194	0.248	0.0604	0.8216
7250	303816.	61904.	2.113	1.494	0.440	0.0019	0.6615
7261	94988.	4125.	22.975	1.657	0.479	0.0464	0.9168
7262	207934.	4857.	14.455	1.151	0.398	0.0449	0.9543
7291	274769.	35882.	0.975	1.150	0.220	0.0237	0.7690
7293	1371813.	45772.	5.924	9.219	0.971	0.0367	0.9354
7294	484288.	38641.	3.930	1.132	0.294	0.0619	0.8522
7295	1298064.	37734.	22.880	0.924	0.260	0.0443	0.9435
7296	184880.	7738.	6.348	0.940	0.308	0.0348	0.9197
7299	1307567.	98031.	1.686	2.323	0.626	0.0453	0.8605
7312	13823.	453.	2.667	1.571	0.348	0.0696	0.9365
7313	75467.	2330.	2.703	1.269	0.468	0.0290	0.9401
7314	52206.	793.	3.620	0.320	0.405	0.0063	0.9701
7315	20804.	5345.	2.173	2.761	0.824	0.0	0.5912
7316	133199.	13814.	0.562	1.480	0.509	0.0485	0.8121
7317	208754.	15932.	0.695	2.081	0.607	0.0398	0.8582
7321	272608.	389831.	1.562	0.993	0.248	0.1402	0.9269
7322	147403.	27325.	2.165	0.405	0.265	0.1106	0.6872
7323	2774172.	78066.	1.451	0.997	0.187	0.1193	0.9453
7324	648897.	8635.	2.115	0.992	0.314	0.0641	0.9737
7325	112342.	0.	1.648	0.623	0.294	0.0	0.0
7327	219645.	14708.	1.759	0.927	0.185	0.0132	0.8745
7328	5654041.	396191.	1.528	1.455	0.300	0.0584	0.8690
7329	631222.	15401.	2.503	2.143	0.371	0.0526	0.9524
7331	208164.	22917.	1.423	1.101	0.251	0.0226	0.8017
7333	341275.	50223.	0.949	1.177	0.347	0.0335	0.7434
7334	1289.	512.	6.274	0.826	0.441	0.0	0.4312

Note: Legend and definitions of variables are given in chapter 8.

Sources: See table C:3.

Table C:5

Basic data for chapter 9. Independent variables in regressions on the changing Swedish commodity group specialization 1964-1970 (1960-1970)

SITC no. (1)	$\Delta Y_W/X_{W,64}$ (2)	ΔP_E \$1000/ metric ton (3)	$\Delta(P_S/P_E)$ (4)	Δh (5)	Δt (6)	$X_{W,64}$ \$1000/ metric ton (7)	$P_{E,64}$ \$1000/ metric ton (8)	$(P_S/P_E)_{64}$ (9)	h_{64} (10)	$\frac{(\Delta C)_{60-70}}{C_{60}}$ (11)	C_{60} 1000 Skr. (12)	C_{64} (13)
6911	1.5970	0.0630	-0.4460	-0.0360	-0.0383	286709.	0.381	1.157	0.341	2.0722	97676.	0.3249
6912	1.8971	0.0710	-0.0810	-0.1150	-0.0300	29903.	2.127	1.234	0.284	4.3246	4260.	0.4656
6913	2.2035	0.2050	-0.2420	-0.4680	0.0	231.	0.514	0.891	0.849	2.9254	67.	0.4761
6921	0.9759	0.1370	1.0010	0.0180	-0.0412	62953.	0.467	1.236	0.378	0.9478	15131.	0.4905
6922	0.9238	0.1610	-0.1210	0.0900	-0.0324	83968.	0.489	0.732	0.411	1.5087	31936.	0.2666
6923	0.3560	0.0870	0.5400	0.1200	-0.0082	46606.	0.569	1.724	0.289	1.9923	3252.	0.7565
6931	1.5139	0.1310	-0.0750	0.1410	-0.0428	103544.	0.502	1.086	0.208	0.8766	14972.	0.7438
6932	-0.1600	0.0470	0.9890	0.9800	-0.0019	28694.	0.170	2.353	0.942	0.2857	637.	0.9556
6933	0.8132	0.0590	2.7130	-0.0250	-0.0335	83153.	0.369	4.523	1.581	1.4876	11151.	0.7087
6934	0.5860	0.0720	7.7710	0.5110	-0.0298	3696.	0.319	0.940	0.508	1.0377	292.	0.7883
6941	0.4107	0.1870	0.3550	-0.3590	-0.0297	72300.	0.233	2.206	1.341	0.8165	8165.	0.7430
6942	1.5865	0.2330	0.1440	-0.3580	-0.0225	173486.	0.743	0.888	1.323	1.3151	28909.	0.6427
6951	0.2087	0.2280	0.3580	-0.0340	-0.0410	33859.	0.728	1.286	0.284	-1.2700	3318.	0.7749
6952	1.1191	1.0680	0.4280	-0.0830	-0.0193	507592.	3.139	1.230	0.582	2.3454	37261.	0.7527
6960	0.6864	1.6010	0.4070	-0.0570	-0.0341	193848.	5.742	1.178	0.370	0.5590	8965.	0.8863
6971	0.2132	0.2230	1.2020	0.0300	-0.0314	129198.	0.945	2.320	0.461	0.0431	7289.	0.8941
6972	1.0022	0.6400	0.6500	-0.1650	-0.0255	112518.	1.535	1.886	0.454	0.7614	12938.	0.7069
6979	1.0463	0.6750	0.0960	-0.0290	-0.0340	22170.	2.078	1.380	0.409	0.5494	2198.	0.7503
6981	1.0055	0.5000	-0.0930	-0.1180	-0.0257	187500.	1.541	0.695	0.323	1.7646	20133.	0.7080
6982	1.0962	0.0980	0.5070	0.0120	-0.0312	8177.	0.809	1.129	0.303	0.8053	3262.	0.3707
6983	0.9469	0.1680	-0.8260	0.0530	-0.0260	74690.	0.703	0.508	0.878	0.6984	8860.	0.7980
6984	1.0907	0.0860	0.6680	0.1210	-0.0658	3185.	0.333	0.697	0.377	0.4221	289.	0.8145
6985	1.3453	1.0210	-0.1140	0.0250	-0.0226	50640.	2.827	1.042	0.332	0.3504	3502.	0.8410
6986	1.4284	0.2200	1.2890	0.0	-0.0309	31152.	0.544	3.774	1.111	1.1750	9879.	0.5189
6988	1.0971	0.3300	-0.4620	-0.0940	-0.0303	103561.	0.783	0.653	0.759	1.2216	16038.	0.6382
6989	1.1716	0.2520	0.7610	-0.1680	-0.0346	308328.	0.810	1.472	0.484	1.6094	58857.	0.5935
8121	1.5113	0.1970	0.4010	-0.1130	-0.0302	84406.	0.387	1.483	0.296	0.7439	25912.	0.2960
8122	0.5198	0.1010	1.3740	0.2420	-0.0391	41582.	0.403	2.037	0.432	1.1225	7641.	0.5908
8123	0.5427	0.2040	1.9110	-0.0900	-0.0220	45018.	0.601	2.531	0.738	1.2503	13789.	0.4574
8124	1.1549	0.1410	0.2760	0.0690	-0.0501	143342.	2.357	1.116	0.282	1.9583	22893.	0.5832
8611	1.9515	11.6280	-0.0760	-0.0790	-0.0103	52841.	47.730	0.681	0.667	3.9083	2192.	0.8614
8612	1.3756	7.1580	1.2590	0.2220	-0.0248	66353.	18.348	1.450	0.358	1.5534	2333.	0.9027

cont.

(C:5 cont.)

(1)	(2)	(3)	(4)	(5)	(6)	(7)	(8)	(9)	(10)	(11)	(12)	(13)
8613	0.9995	7.5960	-0.2050	0.1120	-0.0112	91987.	23.179	0.446	0.574	1.3428	3343.	0.8864
8614	2.4746	-11.0310	1.2950	0.0340	-0.0646	143313.	24.601	3.178	0.667	5.5769	3784.	0.9196
8615	0.7354	5.1410	-0.2070	0.0960	-0.0288	89275.	16.205	1.227	0.413	0.8804	2500.	0.9207
8616	1.1471	1.1910	0.0730	0.0090	-0.0307	145284.	5.361	0.974	0.232	2.1571	2992.	0.9207
8617	1.3995	2.2200	0.2390	0.1170	-0.0356	147723.	12.338	1.487	0.386	2.7179	3623.	0.8773
8618	1.1813	1.4540	2.2740	-0.0170	-0.0270	44643.	7.186	1.143	0.609	1.9439	2710.	0.8527
8619	0.7496	-1.1680	0.3540	0.1210	-0.0249	742751.	11.394	0.806	0.297	1.2328	33053.	0.8741
7111	0.4642	0.1450	0.0340	-0.1370	-0.0470	151184.	1.148	0.797	0.399	1.5492	15527.	0.7702
7112	0.7389	0.4870	0.5400	-0.1200	-0.0590	51189.	1.306	0.790	0.443	0.4886	6003.	0.8555
7113	0.5060	0.4500	0.3850	-0.0200	-0.0784	148810.	3.407	1.269	0.397	1.6647	12054.	0.8388
7115	1.5864	0.3810	0.1750	0.0600	-0.0369	1027070.	2.112	1.047	0.112	1.6904	69273.	0.8145
7116	4.7821	0.7870	-1.9810	0.5320	-0.0870	30544.	6.569	3.836	0.701	3.1876	1940.	0.9222
7117	1.3081	-7.5690	8.2380	-0.4150	0.0025	32247.	12.787	0.800	1.443	0.4509	825.	0.9209
7118	1.0835	1.3980	-0.0770	-0.0040	-0.0538	72426.	1.923	0.841	0.350	0.1818	8112.	0.8197
7120	0.2116	0.2530	0.5400	0.0040	-0.0370	1777681.	1.034	1.282	0.211	1.1601	65295.	0.9036
7141	1.2207	1.5080	-0.3740	0.2870	-0.0275	156595.	6.345	0.949	0.302	1.2176	4334.	0.9408
7142	3.2681	2.5270	-0.3750	-0.0600	-0.0236	325909.	14.574	0.817	0.313	6.7591	11721.	0.8360
7143	0.9790	20.3190	-0.1820	0.1260	-0.0604	372798.	13.854	0.906	0.411	0.0038	9506.	0.9666
7149	4.5664	8.8400	-0.9330	0.3770	-0.0167	295603.	8.482	0.992	0.377	4.2679	11384.	0.8759
7151	0.8490	0.6460	-0.2170	-0.0370	-0.0245	1069264.	2.558	0.841	0.379	1.3938	33330.	0.9103
7152	0.1127	-0.0710	0.6540	0.1150	-0.0490	399186.	0.840	1.749	0.667	1.5103	21189.	0.8828
7171	0.9604	1.1210	-0.2100	0.0690	-0.0144	1168891.	2.710	0.965	0.160	0.9679	11470.	0.9724
7172	0.7495	1.1063	-0.6580	0.0590	-0.0136	50940.	2.770	1.413	0.253	-0.4167	1032.	0.9644
7173	0.5186	1.2440	0.5840	0.0850	-0.0261	280932.	4.077	1.202	0.352	0.5047	12493.	0.9093
7181	0.6767	0.5250	-0.1690	0.0250	-0.0525	354451.	2.379	0.851	0.240	0.5062	24929.	0.8657
7182	0.9041	1.1800	-0.0690	0.0070	-0.0058	386592.	3.023	1.161	0.250	1.6498	17517.	0.8440
7183	0.4769	0.8750	0.1290	0.0430	-0.0345	238369.	1.978	1.429	0.177	0.9835	6863.	0.9178
7184	1.1346	0.1820	-0.0690	-0.0330	0.0344	902425.	1.363	1.063	0.273	1.6663	25191.	0.6383
7185	0.8712	0.3600	0.5020	0.0600	0.0261	324948.	1.541	1.409	0.349	1.1605	14762.	0.6143
7191	1.2055	0.3110	0.3460	0.0070	-0.0348	843072.	1.837	1.324	0.270	1.5269	54483.	0.8347
7192	1.2997	0.4950	-0.0120	0.0090	-0.0343	930879.	2.663	1.199	0.259	1.4356	52158.	0.8377
7193	0.9483	0.2350	0.1880	0.0210	-0.0407	1079748.	1.326	1.025	0.217	1.9538	55656.	0.8383
7194	0.7571	0.5200	-0.2360	-0.0150	-0.0313	46082.	1.796	0.723	0.314	0.1030	2097.	0.8777
7195	1.4180	0.7250	0.6940	-0.0840	-0.0358	431168.	2.722	1.412	0.411	1.3376	21455.	0.8918
7196	1.2204	0.7980	0.3730	0.0130	-0.0292	476602.	3.129	1.287	0.292	2.4323	19641.	0.8755
7197	1.0195	0.3440	-0.2590	0.2150	-0.0301	313592.	2.825	0.876	0.271	0.9160	18937.	0.8622
7198	1.1114	0.5880	-0.0360	-0.0320	-0.0385	884491.	2.718	1.459	0.187	2.7257	30563.	0.8840
7199	1.5118	0.6780	-0.5080	-0.0590	-0.0446	813575.	1.985	0.538	0.303	2.2835	67358.	0.7675
7221	0.7830	0.2700	-0.1110	0.0730	-0.0435	872432.	2.106	1.177	0.187	1.0531	79863.	0.7973

cont.

(C:5 cont.)

(1)	(2)	(3)	(4)	(5)	(6)	(7)	(8)	(9)	(10)	(11)	(12)	(13)
7222	1.5302	2.0720	-0.3490	0.1170	-0.0314	674865.	4.453	1.046	0.190	2.0709	51709.	0.7888
7231	1.2378	0.5900	-0.1210	0.0060	-0.0777	317129.	0.930	1.131	0.308	1.9172	57970.	0.5531
7232	0.6773	-0.2570	0.5300	0.1600	-0.0267	75709.	1.060	0.750	0.620	1.7103	4256.	0.8614
7241	2.9742	1.0140	0.3260	-0.3540	-0.0843	198536.	4.214	1.069	0.546	0.0299	69618.	0.7270
7242	1.5055	2.0460	-0.2260	-0.1370	-0.0607	400592.	6.347	1.912	0.377	0.3199	20808.	0.9067
7249	0.8453	-0.3320	-0.5660	0.2190	-0.0286	1461962.	11.258	0.763	0.346	1.4482	92993.	0.7689
7250	0.9617	0.0060	0.1850	0.0940	-0.0398	194915.	1.269	1.006	0.185	0.2501	33951.	0.6442
7250	1.0227	-0.4230	0.7350	0.0590	-0.0243	148479.	1.665	1.545	0.293	1.4903	12210.	0.7205
7250	0.8803	0.0270	0.3490	-0.0230	-0.0259	131165.	3.127	1.254	0.271	2.1611	7641.	0.7674
7250	1.7419	0.0440	0.7160	0.2300	-0.0056	110803.	2.069	1.267	0.210	2.3337	18569.	0.5854
7261	3.6982	9.0750	0.3410	-0.4510	-0.0320	20218.	13.900	2.359	0.930	5.5476	630.	0.9049
7262	1.7478	3.7120	0.1020	0.2390	-0.0224	75672.	10.743	1.143	0.159	1.0502	2369.	0.9350
7291	1.0243	0.1290	-0.1170	-0.0630	-0.0375	135733.	0.846	1.143	0.283	1.9537	12148.	0.7631
7293	2.3311	1.8950	6.4090	0.1600	-0.0518	411814.	4.029	2.629	0.811	1.6587	17216.	0.9221
7294	1.4297	0.6540	-0.0080	0.0720	-0.0265	199321.	3.276	1.374	0.222	1.9205	13231.	0.8091
7295	1.7603	5.4360	-0.5370	-0.0280	-0.0255	470262.	17.394	0.904	0.288	1.8684	13155.	0.8946
7296	1.3034	0.3770	0.2500	0.0120	-0.0366	80090.	5.971	0.974	0.296	1.8501	2715.	0.9081
7299	1.1026	-0.2920	1.6940	-0.0450	-0.0291	621890.	1.978	1.599	0.671	1.7742	35337.	0.8555
7312	-0.0997	0.4820	0.2890	0.0340	0.0696	15354.	2.185	0.515	0.314	-0.4078	755.	0.9618
7313	-0.5141	0.1300	0.3540	0.1160	-0.0256	155306.	2.573	0.981	0.352	-0.1842	2856.	0.9852
7314	4.0300	1.0000	-0.7120	0.1270	0.0063	10379.	2.620	0.506	0.278	-0.7725	3485.	0.8765
7315	-0.5131	0.4470	1.5040	0.0790	0.0	42727.	1.726	0.053	0.745	0.3521	3953.	0.7530
7316	0.9925	0.0230	0.5930	-0.1390	-0.0016	66851.	0.539	1.373	0.648	0.8167	7604.	0.8249
7317	0.4919	0.1290	1.2930	-0.3690	-0.0227	139927.	0.566	1.452	0.976	0.9713	8082.	0.8837
7321	-0.9270	0.2460	-0.0280	0.1610	0.0065	3736894.	1.316	0.991	0.087	0.7570	221869.	0.8245
7322	0.5975	0.5490	-0.7030	-0.0690	0.0140	92272.	1.616	0.912	0.334	5.3620	4295.	0.7911
7323	1.4919	0.3230	-0.0900	-0.0310	-0.0003	1113264.	1.128	1.082	0.218	0.7450	44737.	0.8676
7324	2.7258	0.2120	-0.2100	0.0700	-0.0437	174165.	1.903	1.296	0.244	6.7236	1118.	0.9832
7325	2.5492	0.3410	-0.6250	0.0610	0.0	31653.	1.307	0.861	0.233	0.0	0.	0.0
7327	0.5062	0.3010	-0.2270	0.0160	-0.0629	145823.	1.458	0.914	0.169	-0.1401	17105.	0.9931
7328	1.8915	0.0680	0.3910	0.0060	-0.0303	1955389.	1.460	1.312	0.294	1.5957	152632.	0.7920
7329	1.8777	0.3450	1.2040	0.2200	-0.0536	219349.	2.158	1.150	0.151	-0.2557	20692.	0.8927
7331	0.7416	0.1750	-0.0320	-0.0470	-0.0782	119524.	1.248	1.125	0.298	1.4336	9417.	0.7585
7333	1.4870	0.0760	0.1530	0.1220	-0.0328	137223.	0.873	1.371	0.225	2.4589	14520.	0.5682
7334	2.7690	2.7530	0.0690	-0.0680	0.0	342.	3.521	1.065	0.509	101.4080	5.	0.7730

Note: Legend and definitions of variables are given in chapter 9.

Sources: See table C:3.

REFERENCES

Åberg, Y. (1969), *Produktion och produktivitet i Sverige 1861-1965*. Industriens Utredninginstitut. Stockholm.

Arrow, K.J., Chenery, H.B., Minhas, B.S. and Solow, R.M. (1961) "Capital-Labor Substitution and Economic Efficiency". *Review of Economics and Statistics* (Aug. 1961).

Bain, J.S. (1956) *Barriers to New Competition*. Harvard University Press. Cambridge.

Balassa, B. (1965a) "Tariff Protection in Industrial Countries: An Evaluation". *Journal of Political Economy*, Vol. LXXIII (Dec. 1965).

- (1965b) "Trade Liberalisation and 'revealed' comparative Advantage". *Manchester School of Economic and Social Studies*, Vol. XXXIII, 1965.

Baldwin, R.E. (1971a) "Determinants of the Commodity Structure of U.S. Trade". *American Economic Review*, Vol. LXI, No. 1 (March 1971).

- (1971b) *Nontariff Distortions of International Trade*. London.

Batra, R.N. (1973) *Studies in the Pure Theory of International Trade*. London.

Batra, R.N. and Casas, F.R. (1973) "Intermediate Products and the Pure Theory of International Trade: A Neo-Heckscher-Ohlin Framework". *American Economic Review*. Vol. LXIII (June 1973).

Bhagwati, J.N. (1965) "The Pure Theory of International Trade: A Survey". In Bhagwati, J.N., *Trade, Tariffs and Growth*. Cambridge 1969. Earlier published without Addendum in *Economic Journal* (March 1964) and in American Economic Association and Royal Economic Society's *Survey of Economic Theory*, Vol. II: Growth and Development. London/New York 1965.

- (1971) "The Generalized Theory of Distortions and Welfare". In Bhagwati, J.N., Jones, R.W., Mundell, R.A. and Vanek, J. (eds) *Trade Balance of Payments and Growth*. Papers in International Economics in Honor of Charles P. Kindleberger.

- (1972) "The Heckscher-Ohlin Theorem in the Multi-Commodity case". *Journal of Political Economy*, Vol. 80, No. 5 (Sept./Oct. 1972).

Bharadwaj, R. (1962) "Factor Proportions and the Structure of Indo-US Trade". *Indian Economic Journal*, Vol. 10 (Oct. 1962).

Bharadwaj, R. and Bhagwati, J. (1967) "Human Capital and the Pattern of Foreign Trade: The Indian Case". *Indian Economic Review*, Vol. 2 (Oct. 1967).

Branson, W.H. (1972) *Factor Inputs, U.S. Trade and the Heckscher-Ohlin Model* (April 1972), mimeo.

Branson, W.H. and Junz, H.B. (1971) "Trends in U.S. Trade and Comparative Advantage". *Brookings Papers on Economic Activity*, No. 2, 1971.

Brown, M. (1969) "Substitution-Composition Effects, Capital Intensity, Uniqueness and Growth". *Economic Journal*, Vol. LXXIX (June 1969).

Buck, T.W. and Atkins, M.M. (1976) "Capital Subsidies and Unemployment Labour, a Regional Production Function Approach". *Regional Studies*, Vol. 10, 1976.

Burenstam Linder, S. (1961) *An Essay on Trade and Transformation*. Uppsala.

Carlsson, B. and Ohlsson, L. (1976) "Structural Determinants of Swedish Foreign Trade, A Test of the Conventional Wisdom". *European Economic Review*, Vol. 7, No. 2, 1976.

Carlsson, B. and Sundström, Å. (1973) *Den svenska importen av industrivaror från låglöneländer*. Industriens Utredningsinstitut, Stockholm.

Caves, R.E. (1974) *International Trade, International Investment, and Imperfect Markets*. Special Papers in International Economics, No. 10, Princeton: International Finance Section, Princeton University.

Caves, R.E. and Khalilzadeh-Shirazi, J. (1976) "International Trade and Industrial Organization: Some Statistical Evidence". *Discussion Paper No. 502*, Harvard Institute of Economic Research, Harvard University, Cambridge, Mass. (Sept. 1976).

Caves, R.E. and Porter, M.E. (1976) "Barriers to Exit", Chapter 3. In Masson, R.T. and Qualls, P.D. (eds) *Essays on Industrial Organisation in Honor of Joe S. Bain*. Cambridge, Mass.

Caves, R.E. and Uekusa, M. (1976) *Industrial Organization in Japan*. The Brookings Institution, Washington, D.C.

Chang, W.W. and Mayer, W. (1973) "Intermediate Goods in a General Equilibrium Trade Model". *International Economic Review*, Vol. 14, No. 2 (June 1973).

Chipman, J.S. (1966) "A Survey of the Theory of International Trade: Part 3, The Modern Theory". *Econometrica*, Vol. XXXIV, No. 1 (Jan. 1966).

Corden, W.M. (1956) "Economic Expansion and International Trade: A Geometric Approach". *Oxford Economic Papers*, Vol. VIII (Sept. 1956).

Du Rietz, G. (1975) *Etablering, nedläggning och industriell tillväxt i Sverige 1954-1970.* Industriens Utredningsinstitut, Stockholm.

- (forthcoming) *Företagsetableringarna i Sverige under efterkrigstiden.* Industriens Utredningsinstitut, Stockholm.

Du Rietz, G. and Hause, J.C. (forthcoming) *Entry, Exit and Micro Dynamics Industry Supply.* Industriens Utredningsinstitut, Stockholm.

Edwards, S.L. (1970) "Transport Cost in British Industry". *Journal of Transport Economics and Policy*, Vol. IV, No. 3 (September 1970).

Esposito, L. and Esposito, F.F. (1971) "Foreign Competition and Domestic Industry Profitability". *Review of Economics and Statistics*, No. 53, (Nov. 1971).

Gort, M. (1962) *Diversification and Integration in American Industry.* Princeton University press. Princeton.

Gray, H.P. (1973) "Two-Way International Trade in Manufactures: A Theoretical Underpinning". *Welwirtschaftliches Archiv*, No. 1 (1973).

Grubel, H.G. (1967) "Intra-Industry Specialization and the Pattern of Trade". *Canadian Journal of Economics and Political Science.* (Aug. 1967).

- (1970) "The Theory of Intra-Industry Trade". In McDougall, I.A. and Snape, R.H. (eds) *Studies in International Economics.* Amsterdam.

Grubel, H.G. and Lloyd, P.J. (1975) *Intra-Industry Trade - The Theory and Measurement of International Trade in Differentiated Products.* London.

Harkness, J. and Kyle, J.F. (1975) *Factors Influencing United States Comparative Advantage.* Economic Department, North-Western University and Exxon Corp., Vol. 5, No. 2 (May 1975).

Heckscher, E.F. (1919) "Utrikeshandelns verkan på inkomstfördelningen". *Ekonomisk Tidskrift*, Part 2. Also published in Ellis, H.S. and Metzler, L.A. (eds) *Readings in the Theory of International Trade.* Philadelphia, 1949.

Herberg, H., Kemp, M.C. and Magee, S.P. (1971) "Factor Market Distortions, the Reversal of Relative Factor Intensities, and the Relation Between Product Prices and Equilibrium Outputs". *Economic Record*, Vol. 47 (Dec. 1971).

Hicks, J.R. (1970) "Elasticity of Substitution Again: Substitutes and Complements". *Oxford Economic Papers*, Vol. 22, No. 3 (Nov. 1970).

Hirsch, S. (1967) *Location of Industry and International Competitiveness*. Oxford.

Hufbauer, G.C. (1970) "The Impact of National Characteristics & Technology on the Commodity Composition of Trade in Manufactured Goods". In Vernon, R. (ed.) *The Technology Factor in International Trade*. NBER, New York.

Hufbauer, G.C. and Chilas, J.G. (1974) "Specialization by Industrial Countries: Extend and Consequences". In Giersch, H. (ed.) *The International Division of Labour*. Problems and Perspectives. Tübingen.

Hufbauer, G.C. and O'Neill, J.P. (1972) "Unit Values of U.S. Machinery Exports". *Journal of International Economics*, Vol. 2, No. 3 (Aug. 1972).

Höglund, B. and Werin, L. (1964) *The Production System of the Swedish Economy. An Input-Output Study*. Industriens Utredningsinstitut, Stockholm.

ILO (1965) *Yearbook of Labour Statistics 1965*.

Jones, R. (1956-57) "Factor Proportions and the Heckscher-Ohlin Theorem". *Review of Economic Studies*, Vol. XXIV, No. 1.

Keesing, D.B, (1965) "Labor Skills and International Trade: Evaluating Many Trade Flows with a Single Measuring Device". *Review of Economics and Statistics*, Vol. XLVII, No. 3 (Aug. 1965).

- (1966) "Labor Skills and Comparative Advantage". *American Economic Review*, Vol. LVI, No. 2 (May 1966).

- (1968a) "Labor Skills and the Structure of Trade in Manufactures". In Kenen, P.B. and Lawrence, R. (eds.) *The Open Economy*. New York.

- (1968b) "The impact of Research and Development on United States Trade". In Kenen, P.B. and Lawrence, R. (eds.) *The Open Economy*. New York.

- (1969) "Scientists and Engineers per 10,000 Workers Required to Produce Eleven Countries' Manufactured Exports in Selected Years from 1913 to 1965 Using Two Sets of Coefficients. Statistical Tables". Presented by Keesing at the Stanford Trade Seminar over "Long Term Changes in the Skill Requirements of Manufacturing Industries". Research Center in Economic Growth. *Memorandum* No. 77 (Aug. 1969). Stanford University, Stanford.

- (1971) "Different Countries' Labor Skill Coefficients and the Skill Intensity of International Trade Flows". *Journal of International Economics*, Vol. 1, No. 4 (Nov. 1971).

Kemp, M.C. and Uekawa, Y. (1972) "Produced Inputs and Their Implications for Trade Theory". *Economic Record* (Dec. 1972).

Kenen, D.B. (1970) "Skills, Human Capital and Comparative Advantage". In Hansen, W.L. (ed.) "Education, Income and Human Capital". NBER, *Studies in Income and Wealth*, No. 35. New York.

Klevmarken, A. (1974) *Industritjänstemännens lönestruktur*. Industriens Utredningsinstitut. Stockholm.

Kravis, I.B. and Lipsey, R.E. (1971a) *Price Competitiveness in World Trade*. NBER, New York.

- (1971b) "International Price Comparisons by Regression Methods". In Griliches (1971) "Price Indexes and Quality Change". *Studies in New Methods of Measurement*. Cambridge, Mass.

Lary, H.B. (1968) *Imports of Manufactures from the Less Developed Countries*. NBER, New York.

Layton, C., Harlow, C. and de Hoghton, C. (1972) *Ten Innovations. An International Study on Technological Development and the Use of Qualified Scientists and Engineers in Ten Industries*. London.

Leontief, W. (1954) "Domestic Production and Foreign Trade: The American Capital Position Re-examined". *Economia Internationale*, Vol. 7, 1954.

- (1956) "Factor Proportions and the Structure of American Trade: Further Theoretical and Empirical Analysis". *Review of Economics and Statistics*, Vol. 38 (Nov. 1956).

- (1964) "An International Comparison of Factor Costs and Factor Use - A Review Article". *American Economic Review* (June 1964).

Lipsey, R.E. and Weiss, M. (1974) "The Structure of Ocean Transport Changes". NBER, *Exploration in Economic Research*, Vol. 1, No. 1, 1974.

Lundberg, L. (1976) *Handelshinder och handelspolitik. Studier av verkningar på svensk ekonomi*. Industriens Utredningsinstitut, Stockholm.

Magee, S.P. (1973) "Factor Market Distortions, Production and Trade: A Survey". *Oxford Economic Papers*, Vol. 25, No. 1 (March 1973),

- (1976) *International Trade and Distortions in Factor Markets*. New York and Basel.

Mayer, W. (1974) "Short-Run and Long-Run Equilibrium for a Small Open Economy". *Journal of Political Economy*, Vol. 82, No. 5 (September/October 1974).

Melvin, J.R. (1968) "Production and Trade with Two Factors and Three Goods". *American Economic Review*, Vol. LVIII, No. 5, (December 1968).

- (1971) "Production Indeterminacy with Three Goods and Two Factors: Reply". *American Economic Review*, Vol. LVI (March 1971).

Minhas, B.S. (1962) "The Homohypallagic Production Function, Factor-Intensity Reversals and the Heckscher-Ohlin Theorem". *Journal of Political Economy*, Vol. LXX, No. 2 (April 1962).

Moody, C.E., Jr (1974) "The Measurement of Capital Services by Electrical Energy". *Oxford Bulletin in Economics and Statistics*. 36, no. 1. (Feb. 1974).

Nadiri, M.I. (1970) "Some Approaches to the Theory and Measurement of Total Factor Productivity: A Survey". *Journal of Economic Literature*, Vol. VIII, No. 4 (December 1970).

Norstedt, J.-P. (1972) "Annorlunda principer för lönesättning i USA, Yrkesarbetet premieras bättre än i Sverige", *Arbetsgivaren*, no. 40, 1972.

Norström, G. (1974) Transportgeografiska studier i svensk utrikeshandel. Ekonomiska Forskningsinstitutet vid Handelshögskolan i Stockholm.

OECD (1970) *Gaps in Technology*, Analytical Report. Comparisons between Member Countries in Education, Research and Development. Technological Innovation and International Economic Exchanges. Paris.

Ohlsson, L. (1969) *Utrikeshandeln och den ekonomiska tillväxten i Sverige* 1871-1966. Industriens Utredningsinstitut. Stockholm.

- (1973) *Metallmanufakturindustrin - Produktionsförutsättningar och specialisering i internationell jämförelse.* Industriens Utredningsinstitut. Stockholm.

- (1974a) *Factor Abundance, Trade Specialisation and Lags in Adjustment* - A Case Study of Swedish Engineering Trade, mimeo. Industriens Utredningsinstitut, Stockholm. Revised version published in *Journal of Political Economy*, No. 2. 1977.

- (1974b) "On Unit Prices and Their Use in the Analysis of the International Specialisation Pattern with Heterogeneous Industries". *Journal of International Economics*, Vol. 4, No. 3, 1974.

- (1975a) "Specialisation Tendencies in Swedish Trade and Production of Fabricated Metal Products in the 1960's". *Swedish Journal of Economics*, Vol. 77, No. 3. (Sept. 1975).

- (1975b) *Trends and Interdependencies in Engineering Trade Patterns of Industrial Countries 1964-70*, mimeo from Industriens Utredningsinstitut presented at Second Conference on Economics of European Industrial Structure in Nijenrode, Holland (April 1975).

Pagoulatis, E. and Sorenson, R. (1976a) "Domestic Market Structure and International Trade: An Empirical Analysis", *Quarterly Review of Economics and Business*, No. 16 (Spring 1976),

- (1976b) "International Trade, International Investment and Industrial Profitability of U.S. Manufacturing", *Southern Economic Journal*, No. 42 (January 1976).

Paues, W. (1958) *Europamarknaden och företaget*, Studieföfbundet Näringsliv och Samhälle. Stockholm.

Ray, A. (1972) "Traded and Non-Traded Intermediate Inputs and the Rybczynski Theorem", *International Economic Review*, Vol. 13, No. 3 (October 1972).

Rosefielde, S. (1973) *Soviet International Trade in Heckscher-Ohlin Perspective. An Input-Output Study.* Lexington, Mass.

Roskamp, K.W. and McMeekin, G.C. (1968) "Factor Proportions, Human Capital and Foreign Trade: The Case of West Germany Reconsidered", *Quarterly Journal of Economics*, Vol. LXXXII, No. 1 (February 1968).

Rybczynski, T.M. (1955), "Factor Endowment and Relative Commodity Prices". *Economica*, Vol. 22 (November 1955).

SACO (1971) *Inkomstpolitiska fakta och riktlinjer.* Stockholm.

SAF (1967 and 1971) *Direct and Total Wage Cost for Workers.*

- (1973 *Löneskillnader i några länder.* Svenska Arbetsgivareföreningen. Stockholm.

Samuelson, P.A., (1953-54) "Prices of Factors and Goods in General Equilibrium". *Review of Economic Studies*, Vol. 21.

Sato, R. and Koizumi, T. (1973) "On the Elasticities of Substitution and Complementarity". *Oxford Economic Papers*, Vol. 25, No. 1, March 1973.

Stewart, D.B., 1971,"Production Indeterminacy with Three Goods and Two Factors: A Comment on the Pattern of Trade". *American Economic Review*, Vol. 61, March 1971.

Stolper, W.F. and Roskamp, K.W., "Input-Output Table for East Germany with Applications to Foreign Trade". *Bulletin of the Oxford University Institute of Statistics*, Vol. 23, Nov. 1961.

Stolper, W.F. and Samuelson, P.A., (1941). "Protection and Real Wages", *Review of Economic Studies*, Vol. 9, (Nov. 1941).

Tatemoto, M. and Ichimura, S., (1959) "Factor Proportions and Foreign Trade: The Case of Japan". *Review of Economics and Statistics*, Vol. 41 (Nov. 1959).

Travis, W.P., (1972) "Production, Trade and Protection When There Are Many Commodities and Two Factors". *American Economic Review*, Vol. 62 (March 1972).

Vernon, R., (1966) "International Investment and International Trade in the Product Cycle". *Quarterly Journal of Economics*, (May 1966).

Wahl, D.F., (1961) "Capital and Labour Requirements for Canada's Foreign Trade". *Canadian Journal of Economics and Political Science*, Vol. 27 (Aug. 1961).

Wells, L.T., (1972) "International Trade: The Product Life Cycle Approach" Publicerad i Wells, L.T. (ed.), *The Product Life Cycle and International Trade*, Boston.

White, L.J. (1974) "Industrial Organization and International Trade: Some Theoretical Considerations", *American Economic Review*, 64 (December 1974).

Wijkman, P.M. (1974) Book review of Ohlsson, L. in *Statistisk Tidskrift*, (1974).

Vision, (1974) The European Business Magazine, No. 44/45 (1974).

Yeung, P. and Tsang, H., (1972) "Generalized Production Function and Factor-Intensity Crossovers: An Empirical Analysis", *Economic Record*, (Sept. 1972).

INDEX